INTERNATIONAL SERIES ON
MATERIALS SCIENCE AND TECHNOLOGY
EDITOR: PROFESSOR W. S. OWEN

VOLUME 12

Introduction to the Properties of
CRYSTAL SURFACES

Introduction to the Properties of

CRYSTAL SURFACES

J. M. BLAKELY

Professor of Materials Science and Engineering
Cornell University

PERGAMON PRESS

OXFORD · NEW YORK · TORONTO
SYDNEY · BRAUNSCHWEIG

PHYSICS

Pergamon Press Ltd., Headington Hill Hall, Oxford
Pergamon Press Inc., Maxwell House, Fairview Park, Elmsford,
New York 10523
Pergamon of Canada Ltd., 207 Queen's Quay West, Toronto 1
Pergamon Press (Aust.) Pty. Ltd., 19a Boundary Street,
Rushcutters Bay, N.S.W. 2011, Australia
Vieweg & Sohn GmbH, Burgplatz 1, Braunschweig

First edition 1973

Library of Congress Cataloging in Publication Data

Blakely, John McDonald.
Introduction to the properties of crystal surfaces.

(International series on materials
science and technology, v. 12)
Bibliography: p.
1. Surface chemistry. 2. Crystals. I. Title.
II. Title: Crystal surfaces.
QD506.B56 1973   541'.3453   73-6996
ISBN 0-08-017641-0

Printed in Hungary

CONTENTS

LIST OF TABLES

PREFACE

SURFACE science is concerned with the structure and properties of the transition region between phases. It ranges over many of the traditional disciplines and is an important research area in biology, metallurgy, and solid state physics. In this text we will be concerned with crystal surfaces. The theoretical and experimental techniques used in the study of crystal surfaces are in many cases extensions of the methods which have been so successful in developing an understanding of the bulk properties of materials. In treating most bulk properties the presence of the external surface is either ignored completely or is adequately accounted for by the introduction of convenient, although perhaps unrealistic, boundary conditions. There are, however, a great many phenomena which can only be described in a satisfactory way by considering the details of the atomic and electronic structure in the surface region. Examples of such phenomena are electron emission, adsorption and oxidation, adhesion, friction, nucleation and epitaxial growth, and heterogeneous catalysis. It is hoped that this text will provide some of the necessary background for further detailed study of these various topics.

For many processes it is possible to develop both a phenomenological or macroscopic description and a microscopic one. The two are, of course, complementary. The macroscopic description introduces parameters which can only be determined by experiment or from a detailed microscopic theory, but the use of macroscopic thermodynamics in surface science has proved extremely useful in formulating descriptions of a number of properties and phenomena. Most aspects of the thermodynamic description of surfaces are based on the classic work of Gibbs.

Our experience with physical properties of bulk crystals has emphasized the strong dependence of properties on structure, and there is considerable evidence to suggest that similar types of correlation are to be expected for surface properties. Near the surface of a crystal we may expect different arrangements of the atoms or ion cores and departure of the electronic distribution from its strictly three-dimensionally periodic arrangement. New types of lattice imperfections will exist, and there will generally be changes in the density of the usual bulk defects as the surface is approached. The depth to which the surface influences the crystal structure will depend, of course, on the type of crystal and particularly on the degree of delocalization of the valence or bonding electrons. As with other defects in crystals, the free surface may have associated with it additional electronic states and lattice vibrational modes. These can be influenced by the binding of foreign atoms or ions to the surface. The electronic surface states are of prime importance in determining the electrical transport properties of semiconductor surfaces and the performance of some semiconductor devices. This special topic of semiconductor surface behavior is treated in detail in the fairly recent books by Many, Goldstein, and Grover, and by Frankl: it is discussed only very briefly in the present text.

The last decade has been a boom period for the development of experimental techniques in surface studies. Facilities are now quite commonly available for the production and maintenance of uncontaminated surfaces, and techniques for studying the atomic structure of the last few layers are being developed. A great many new features have already been discovered and it is likely that there will be rapid developments in this field.

The present text is intended to serve as an introduction to the study of surface phenomena and is aimed at senior undergraduates in a materials science type of curriculum or those beginning research work in the field or associated areas. The recommended background for this text and for the additional reading material is a knowledge of thermodynamics, atomic physics, and solid state physics at the levels normally taught in engineering or science degree programs. The material covered is, of course, biased to some extent by the au-

thor's own research interests. Although some effort has been made
to indicate the more significant recent advances in a number of
areas, no attempt has been made to give complete reviews or to establish priorities as it is felt that this is not of immediate importance to
the student. Much of the material in this text may be found in the
numerous papers and reviews collected and published as conference
proceedings. It is felt that having covered the present text the student
will be better equipped for a critical reading of these contributions.

The reading material listed at the end of each chapter is intended as
a guide to the literature and should be a useful reference source.
The articles cited are mostly books and recent reviews.

I would like to acknowledge the suggestions and advice of a number
of colleagues during the preparation of the manuscript. I am indebted
to James C. Shelton for carefully reading and offering many useful
comments on the entire manuscript. Also of special value have been
the detailed criticisms and comments from S. Danyluk and H. Patil
of the Materials Science Department at Cornell, D. Tabor and A. E.
Lee of the Cavendish Laboratory, Cambridge, where much of the
manuscript was written, and Professors W. S. Owen and J. E. Hilliard
of Northwestern University. I am also deeply indebted to the large
number of researchers who kindly supplied photographs and draw-
ings from their original work; individual acknowledgements accompany the figures. I would also like to acknowledge a generous fellow-
ship from the Guggenheim Foundation,

I am very grateful to Mrs. Karen Murphy and Mrs. Nellie Zorc
for their help in typing the manuscript and to my wife for her constant support.

Cornell University, Ithaca, NY

CHAPTER 1

THERMODYNAMICS OF SURFACES

1.1. INTRODUCTION

Equilibrium thermodynamics is a subject based on three basic postulates, or laws, in which one derives relationships among the various state functions such as internal energy E, entropy S, enthalpy H, Helmholtz free energy F, Gibbs free energy G, etc., and the state parameters temperature T, pressure P, volume V, mole fraction x_i of component i, etc. One can derive a number of useful relationships which apply to surfaces.

The extensive thermodynamic properties of a solid will include contributions which depend on the area (and perhaps the shape) of its surface. These are normally (and justifiably) neglected in treating properties of the bulk solid but are of considerable interest for our present purpose. There are different ways in which the thermodynamic properties of surfaces can be defined. For example, if we consider an interface separating two otherwise homogeneous phases α and β, the surface thermodynamic functions may be defined in terms of a *surface phase* or by introducing the concept of a *dividing surface*. In the first method the system is considered to be one in which there are three phases present—the two bulk phases and a surface phase: the boundaries of the surface phase are somewhat arbitrary and are usually chosen to be at locations at which the properties are no longer varying significantly with position. The surface phase then has a finite volume and may be assigned thermodynamic properties in the normal way. With the method involving a single dividing surface,

1

the surface contributions to the thermodynamic functions are defined as the excesses over the values that would obtain if the bulk phases retained their properties constant up to an imaginary surface separating the two phases. We adopt the latter procedure here and in the present chapter we consider only one-component crystals.

In discussions of thermodynamic properties of any system, the free energy functions generally play an important role since they can be used for developing convenient criteria for equilibrium. Surface tension plays a similar role in thermodynamic treatments of surface properties, and we shall devote most of the present chapter to a discussion of surface tension in solids. Much of the development follows that for the case of simple liquids but with important modifications arising from the dependence of solid surface properties on crystallographic orientation and from the relatively low mobility of atoms in the solid state. The first of these is important in determining the equilibrium shapes of small crystals and the stability of planar surfaces with normals along particular crystallographic directions. The second leads to a possible distinction between surface stress and surface tension, whereas for liquids the two quantities are always numerically equal. In this chapter we discuss also the equilibrium configuration at the intersection of interfaces and the effects of curvature of crystalline surfaces.

1.2. ONE-COMPONENT SYSTEMS

Consider an interface between two phases α and β (solid–vapor, solid–liquid, solid–solid, etc.) in a one-component system. We will suppose the phases α and β to extend sufficiently far from the interface that we may characterize them by their bulk concentrations C^α, C^β (moles/unit volume) as indicated in Fig. 1.1. For the purpose of defining surface quantities we imagine a surface (DS, Fig. 1.1) to be constructed separating the two phases and coinciding approximately with the transition region between α and β. We can then define any extensive property P of the surface or interface by an equation of the type

$$P_{\text{total}} = P^\alpha + P^\beta + P^s, \tag{1.1}$$

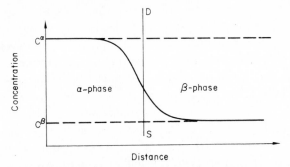

FIG. 1.1. Variation of the concentration of a particular component across the interface between two phases α and β. C^α and C^β are the concentrations of that component in the two phases at large distances from the interface. DS is the dividing surface.

where P^α and P^β are the values of the extensive quantity for the α and β phases respectively if they continued *homogeneous* up to the dividing surface. P_{total} is the value of the quantity for the entire system and P^s the value of P to be associated with the surface. Defined in this way it is clear that P^s is to be regarded as the *excess* value of P for the real system compared to that of the imaginary system consisting of two homogeneous phases with an ideally discontinuous change in properties at a particular mathematical surface. As examples we may write

$$S^s = S_{\text{total}} - (S^\alpha + S^\beta),$$
$$E^s = E_{\text{total}} - (E^\alpha + E^\beta), \qquad (1.2)$$
$$F^s = F_{\text{total}} - (F^\alpha + F^\beta)$$

for the surface excess entropy, internal energy, and Helmholtz free energy respectively. Also a quantity which is of interest in connection with adsorption and segregation phenomena is the excess amount of material or excess number of moles n^s to be associated with the interface. This is defined as

$$n^s = n_{\text{total}} - (n^\alpha + n^\beta)$$
$$= n_{\text{total}} - (C^\alpha V^\alpha + C^\beta V^\beta), \qquad (1.3)$$

where n_{total} is the total number of moles in the entire system and n^α and n^β refer to the α and β phases respectively; V^α and V^β are the volumes of the two phases measured to the dividing surface and C^α and C^β the concentrations (moles per unit volume) in the homogeneous phases. For simplicity we may specialize to planar interfaces with no variations in properties parallel to the interface. (We should expect the results derived here to remain valid for nonplanar interfaces provided the radius of curvature is significantly greater than the width of the transition region.) The quantities in eqns. (1.1)–(1.3) are then defined[†] with respect to a cylinder of unit cross-sectional area perpendicular to the interface and extending into the phases α and β. The excess number of moles per unit area is generally denoted by Γ and often referred to simply as the surface excess.

It is obvious that the value of any surface property defined by the above equations will depend upon the choice of the location for the dividing surface. We will consider below a particularly convenient choice.

The *surface tension* γ may be defined as the reversible work involved in creating unit area of *new* surface at constant temperature, volume, and total number of moles,

$$\gamma = \lim_{dA \to 0} \frac{dw}{dA}, \tag{1.4}$$

where dw is the amount of work associated with the increment dA in area.

Since at constant temperature and volume the work done is equal to the change in Helmholtz free energy of the whole system

$$\gamma \, dA = dF_{total} = d(F^\alpha + F^\beta) + dF^s$$
$$= \mu \, d(n^\alpha + n^\beta) + dF^s$$
$$= -\Gamma \mu \, dA + f^s \, dA$$

or

$$\gamma = f^s - \Gamma \mu, \tag{1.5}$$

[†] To denote the values of thermodynamic functions per unit area of surface we use lower-case letters, e.g. e^s, s^s, f^s represent the internal energy, entropy, and Helmholtz free energy per unit area.

where μ is the chemical potential and f^s is the value of F^s per unit area or, in other words, the *specific surface free energy*. We may note from (1.5) that γ is the surface density of the quantity $(F-G)$ (usually referred to as the Ω potential), a relationship which was used by Gibbs for the definition of γ. We shall use this definition in Chapter 2 in connection with the composition variation near the surface of a binary solid.

Since Γ and f^s were defined with respect to an arbitrary dividing surface their individual values will depend on the choice of this surface. However, γ as defined by eqn. (1.4) is clearly independent of the choice so that the surface tension has a unique value for any particular interface. Thus in general the surface tension and specific surface free energy are not equal. This fact is of more importance in connection with multicomponent systems since for a one-component system it will in general be possible to choose the dividing surface such that $\Gamma = 0$. This is, in fact, the conventional choice for a single-component system, and thus the surface tension and specific free energy may be identified in this case.

Surface tension is a quantity which is directly measurable (see Chapter 3), and evidently by measuring its temperature dependence we may extract values of the surface (internal) energy and entropy, quantities which will reflect the differences in binding and vibrational motion of the surface atoms. Since

$$f^s = e^s - Ts^s, \quad \text{where} \quad s^s = -\left(\frac{\partial f^s}{\partial T}\right)_{v,\Gamma},$$

we obtain (for one component)

$$e^s = \gamma - T\left(\frac{\partial \gamma}{\partial T}\right)_v. \tag{1.6}$$

The experimental data on liquids indicate in general that $(\partial \gamma / \partial T)$ is negative which in turn implies a positive excess surface entropy. The temperature coefficient of γ for solids also seems to be negative although the data on γ as a function of T for solids are still rather sparse. Such information would, however, be extremely valuable for testing

B: IPCS: 2,

calculations, based on microscopic models, of the internal energy and entropy contributions to the surface tension. The empirical relationship

$$\gamma = \beta(T - T_c) \tag{1.7}$$

for the variation of γ with T known as the Eötvös law is found to hold quite well for many simple liquids. Here T_c is the critical temperature and β a constant characteristic of the material. In the absence of a better model the surface tension of solids is usually also taken to vary linearly with temperature. Some experimental data for copper through the equilibrium melting temperature T_m are shown in Fig. 1.2. The scatter is clearly too great to allow a comparison of the internal energy and entropy contribution to the surface tensions of the phases, but the discontinuity of about 25% in γ at the melting point

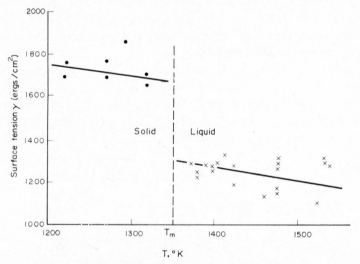

FIG. 1.2. Variation of surface tension of copper with temperature through the melting point T_m. Although there is considerable scatter in the data, the discontinuity in γ at the melting point is quite clear. (Data for the solid from H. Udin, A. J. Shaler, and J. Wulff, *Trans. AIME* **185**, 186 (1949). (Data for the liquid from tabulations in P. Kozakevitch, in *Liquids: Structure, Properties, Solid Interaction*, ed. T. J. Hughel, Elsevier, 1965, and in D. A. Belforti and M. P. Lopie, *Trans. AIME* **227**, 20 (1963).)

seems to be quite well established. In general the ratio of the surface tension of the solid to that of the liquid at the melting point appears to be about 1.1–1.3 from the available data on metals and because of the greater availability of experimental surface tensions for liquids the value for the solid is often extrapolated from the liquid data; an estimate of the solid surface tension at a temperature below T_m may be obtained by assuming

$$\frac{1}{\gamma(T_m)}\ \frac{\partial \gamma}{\partial T} \simeq -0.2 \times 10^{-4}/^{\circ}\text{K}$$

for the common metals.

Some numerical values of surface tension for nominally pure solids are given in Table 3.2 (p. 64). Although some of these results may have been influenced by small amounts of impurities, the data suggest that for common metals such as copper, nickel, etc., the value of γ is about $1-2 \times 10^3$ ergs/cm² or approximately 0.4–1.0 eV for each atom in an atomic plane parallel to the dividing surface.

1.3. SURFACE TENSION AND SURFACE STRESS

There exists considerable confusion in the literature on solid surfaces about the meaning of the quantities surface tension and surface stress. The difficulty arises partly from the use of different terminology by different authors, but in many cases it is associated with the fact that most of the earlier work on surfaces was concerned with liquid–vapor interfaces for which the distinction is unimportant, and in fact the two quantities are numerically equal in that case. In general, surface stress is a tensor quantity, whereas surface tension as defined by eqn. (1.4) or (1.5) is evidently a scalar. (In section 1.4 we shall in fact see that γ varies with the surface orientation and is therefore a scalar function of the unit vector along the surface normal.) Perhaps the best discussion of the meaning of surface tension and stress in crystalline solids is that given in a now well known series of publications by Herring, and we shall outline here the main features.

2*

Surface tension corresponds to the work to create unit area of new surface whereas surface stress is involved in computing the work involved in deforming a surface. As we shall see, the two quantities will be numerically equal when atomic mobilities are sufficiently high to preserve the microscopic configuration of the surface following the deformation. More quantitatively, consider an arbitrary change dA in the area A of the surface; this may be expressed in terms of a change of the strain tensor ε_{ij} describing the surface plane by $\Delta\varepsilon_{ij}$ defined by

$$dA = A\sum \Delta\varepsilon_{ij}\,\delta_{ij} \qquad (i, j, = 1, 2). \qquad (1.8)$$

The amount of work required for such a deformation may be written (to first order) by defining a surface stress tensor g_{ij} such that

$$dw = A\sum g_{ij}\,\Delta\varepsilon_{ij} \qquad (i, j = 1, 2). \qquad (1.9)$$

The amount of work required is also equal to the change in the quantity γA, so that

$$dw = d(\gamma A) = \gamma\,dA + A\,d\gamma$$
$$= \gamma A \sum_{i,j} \Delta\varepsilon_{ij}\,\delta_{ij} + A\sum_{i,j} \left(\frac{\partial\gamma}{\partial\varepsilon_{ij}}\right) \Delta\varepsilon_{ij}. \qquad (1.10)$$

Hence equating the right hand sides of eqns. (1.9) and (1.10) and noting that the resulting equality is true for any arbitrary additional strain component $\Delta\varepsilon_{ij}$,

$$g_{ij} = \gamma\,\delta_{ij} + \left(\frac{\partial\gamma}{\partial\varepsilon_{ij}}\right) \qquad (1.11)$$

as the desired relationship between surface stress g_{ij} and surface tension γ. Thus we see that, in general, surface stress and surface tension will be numerically equal only if γ is unaffected by the deformation. This is evidently true in the case of a liquid where atomic mobilities are high and there is no long range correlation in atomic positions. The surface stress will be isotropic with zero shear components (i.e.

$g_{11} = g_{22}$, and $g_{12} = g_{21} = 0$), so that it may be characterized by a single quantity g where

$$g = \gamma. \tag{1.12}$$

For a solid, due to the long range correlation in atomic positions and low atomic mobilities, it may not be possible, in any reasonable experimental time, to keep constant the local configuration around any particular atom in the surface region where the deformation of the surface area is performed. Hence in this case γ will be altered, i.e. $(\partial \gamma / \partial \varepsilon_{ij}) \neq 0$ in general for a crystalline surface. Since γ is almost invariably[†] positive as evidenced by the fact that the surface areas of condensed phases do not spontaneously expand, the surface stress of liquids is always positive or tensile. This produces, e.g. in a small liquid droplet, a tendency to contract, resulting in an increase in density or a decrease in mean atomic volume. On the other hand, the surface stress of crystalline solids can be either tensile (positive) or compressive (negative) depending on the magnitude and sign of $(\partial \gamma / \partial \varepsilon_{ij})$. Thus small crystalline particles of some materials may be expected to show some increase of atomic volume relative to bulk material, whereas those of other materials will have decreased mean atomic volume.

We should note in passing that the existence of a surface stress does not in itself imply that the arrangement of atoms in the surface region is different from that in the interior of the crystal. The condition for the existence of such rearrangements in a crystal at equilibrium is that they correspond to a minimization of the total free energy or of the total internal energy at $T = 0°K$. It is also worth commenting that in a solid even at high temperatures we would generally not expect the surface stress tensor to be zero although this is obviously a possibility. Mechanisms whereby the surface stress state of a solid can be altered can be devised, but it is considerably more difficult to predict whether they will be energetically favorable. Figure 1.3

[†] The surface tension associated with the interface between normal and superconducting regions in type II superconductors may effectively be negative.

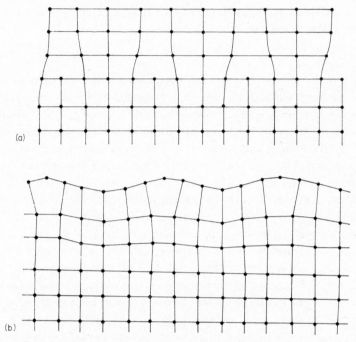

FIG. 1.3. Illustration of how the state of compression or tension in the surface layers can be altered by suitable arrays of dislocations. In (a) the array of dislocations allows the mean spacing in the surface layers to be greater than that in the bulk; (b) illustrates a possible "rumpling" effect which effectively decreases the surface spacing.

indicates examples of possible surface configurations, involving arrays of dislocations, which would lead to changes in the surface stress from that characteristic of an ideal plane.

1.4. VARIATION OF SURFACE TENSION WITH ORIENTATION

A study of the arrangements of atoms in different planes of a crystal will immediately suggest that most properties associated with solid surfaces will vary with the crystallographic orientation of the surface plane.[†] This is confirmed by experimental studies of such diverse phenomena as chemical reactions between surfaces and solutions, electron emission, and surface atomic diffusion. Surface tension is also expected to vary with orientation since the binding energies and vibrational modes of the surface atoms will depend on the local atomic arrangement. If this variation of γ with orientation is sufficiently marked, the equilibrium shapes of crystals will be polyhedral and planar surfaces of certain orientations may be unstable with respect to a spontaneous decomposition into a surface composed of segments of two or more other orientations even although this process involves a net increase in the real surface area. We will consider here the connection between crystal morphology and the variation of γ with orientation. This problem is of considerable interest in heterogeneous catalysis by small metallic particles since catalytic efficiency may be a strong function of surface orientation. In a later section we return to a discussion of experimental techniques of determining γ.

Consider a one-component system (for which we may use the terms surface tension and specific surface free energy interchangeably) in

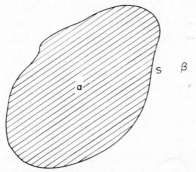

FIG. 1.4. Particle of α-phase separated from the β-phase by an interface (or dividing surface) S, e.g. a crystalline particle in contact with its vapor. The equilibrium shape of S is that given by the inner envelope of the γ-plot of Fig. 1.5.

---

† An excellent and extensive array of ball models of crystalline surfaces is given in the book by J. F. Nicholas, *An Atlas of Models of Crystal Surfaces*, Gordon & Breach, 1965. See also Chapter 4.

which there are two phases α and β separated by an interface (Fig. 1.4). The problem we wish to focus on is that of determining the equilibrium shape of the interface. At fixed temperature and total volume the condition of equilibrium is one of minimum Helmholtz free energy. We will suppose the phases α and β to have their equilibrium volumes and will consider changes only with respect to alterations in the boundary shape. With these restrictions the equilibrium condition is that

$$\int_S \gamma(\hat{\mathbf{n}}) \, dA = \text{minimum}, \qquad (1.13)$$

when $\gamma(\hat{\mathbf{n}})$ means the surface tension of a surface whose orientation is denoted by the unit vector $\hat{\mathbf{n}}$ along the normal and the integral is taken over the interface S. We are, of course, neglecting all external gravitational, electrical, or magnetic fields although the external gravitational field in particular may be of importance in real situations.[†]

For interfaces between simple liquids and gases where γ is independent of orientation, it is clear that the condition of equilibrium is simply one of minimum surface area so that the equilibrium interface shape is spherical. There is some experimental evidence that in certain crystalline substances the crystal–vapor interface approaches this condition at elevated temperatures. However, when there is appreciable variation of γ with orientation the resulting equilibrium shape will be polyhedral with surfaces of low γ being preferentially exposed. There is, of course, no way in which $\gamma(\hat{\mathbf{n}})$ can be deduced from thermodynamics; it must be either measured or computed from a microscopic model. We will suppose for our present purpose that $\gamma(\hat{\mathbf{n}})$ has been so determined and ask, Given $\gamma(\hat{\mathbf{n}})$ what is the equilibrium shape?

The variation of γ is most conveniently represented by a polar diagram, called the *Wulff plot*, in which the radius vector represents the orientation of the surface (i.e. the direction of the surface normal $\hat{\mathbf{n}}$) and the magnitude of the surface tension, i.e. it is the plot $\mathbf{r} = \gamma(\hat{\mathbf{n}})\hat{\mathbf{n}}$. A two-dimensional section perpendicular to a [100] direction

---

[†] The distortion of small droplets produced by the gravitational field can n fact be used in determining the surface tension of liquids as, for example, in the sessile drop method (see, for example, A. W. Adamson, *Physical Chemistry of Surfaces*, Interscience, 1967).

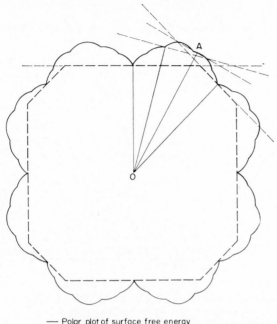

— Polar plot of surface free energy
--- Samples of planes normal to radius vectors of this plot
—— Equilibrium polyhedron

FIG. 1.5. Two-dimensional section of a polar plot of surface tension (Wulff plot) for a cubic crystal in which the vector from the origin to any point on the plot represents the direction of the normal to a particular plane and the magnitude of the surface tension for that particular plane. The equilibrium shape of a crystal can be derived from its Wulff plot; it is the inner envelope of Wulff planes. Surfaces with orientations such as A, which are not present in the equilibrium or Wulff shape, may be metastable with respect to faceting (see text). (After C. Herring in *Structure and Properties of Solid Surfaces*, ed. R. Gomer and C. S. Smith, University of Chicago Press, 1953, Chapter 1.)

of the Wulff plot of a cubic crystal is shown schematically in Fig. 1.5. The Wulff plot will have symmetry properties which are the same as those of the crystal and for a cubic crystal: for example, the entire plot may be specified by considering only one octant of the diagram. Thus the γ-plot may then alternatively be represented by lines of constant surface tension in the stereographic triangle as indicated in Fig. 1.6, and

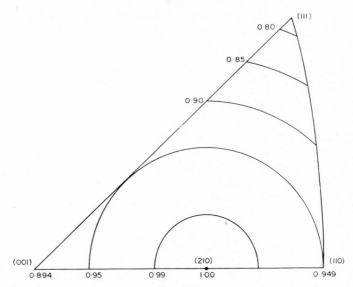

FIG. 1.6. Representation of the γ-plot of a cubic crystal in the stereographic triangle. The lines are contours of constant surface tension and are normalized with respect to the value for the (210) surface. This particular plot is a calculated one for a face-centered cubic crystal and is based on nearest neighbor interactions (see section 5.3). (Courtesy W. Winterbottom and N. A. Gjostein.)

this is generally the most convenient method for presenting experimental data. Figures 1.5 and 1.6 have been drawn to indicate that certain orientations may correspond to local minima in surface tension. These minima have generally been referred to in the literature as *cusps* although they will not, in general, strictly satisfy the mathematical definition of a cusp which would require the slope of the γ-plot to become infinite. The surfaces corresponding to the cusp orientations are termed *singular*.

If we consider any point P on the diagram the plane through P perpendicular to the radius vector OP is referred to as a *Wulff plane* and is obviously parallel to the crystal surface to which the point P refers. The equilibrium shape of the crystal is determined from the γ-plot by a procedure known as the *Wulff construction* which consists of finding the inner envelope of the Wulff planes for all possible directions. This is shown schematically in Fig. 1.5. Clearly the Wulff construction gives the correct result for a liquid. The proof of the correctness of this procedure for crystalline materials has been given by a number of authors with different degrees of generality.[†] We may note that although it is possible by this construction to go uniquely from the Wulff plot to the equilibrium shape, the reverse procedure is clearly not possible although the ratios of the surface tension for various pairs of orientations which are present in the final shape could be determined from the equilibrium shape.

At elevated temperatures where atomic transport rates become appreciable, a phenomenon known as *faceting* frequently occurs. This consists of the break-up of an initially flat surface into a hill and valley structure which is made up of portions of two or more other orientations one of which is generally a low index plane (see Figs. 1.8 and 1.9). The exposed flat areas of this low index plane are called *facets*. Since the effect may have a significant influence on many surface properties such as average electronic work function, catalytic activity, or oxidation rate, etc., it is worth while to consider it in some detail and in particular to investigate the criteria for the stability of planar surfaces of a given orientation. These criteria can most readily be developed with the aid of the Wulff plot and, in fact, we can assert immediately that a flat surface which is of an orientation present in the equilibrium shape will be stable with respect to faceting; surfaces of other orientations will exhibit faceting. Thus if an experimental Wulff plot were available for the temperature range and environment of interest it would be possible to predict whether or not a particular surface would be stable and also to determine the new surfaces which would be exposed if faceting should occur. Extensive data on Wulff plots are not yet available, and for this reason stability criteria have been developed which involve only a small segment of the polar plot of γ. If the surface with which we are dealing has an

---

[†] See, for example, the article The kinetic and thermodynamic properties of surfaces, by J. P. Hirth in *Energetics in Metallurgical Phenomena* (ed. W. M. Mueller), Gordon & Breach, 1965.

orientation close to that of some low index plane S we may be interested in the stability of our surface with respect to exposing segments of S. This question can most readily be answered using the so-called "Herring tangent sphere criterion". If we consider a section of the γ-plot passing through some singular orientation S and the surface of interest A (Fig. 1.7), the condition for stability of A with respect to formation of portions of S is that the sphere drawn through A and through O, the γ-plot origin, should lie inside the γ-plot between

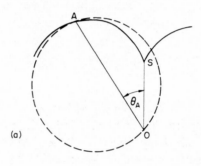

(a)

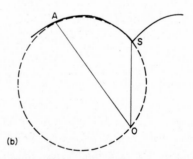

(b)

Fig. 1.7. Tangent sphere criterion for stability of a particular surface with respect to faceting. OA represents the surface tension and orientation of a surface A close to some surface S at which a γ-plot cusp occurs. A is unstable with respect to faceting if the sphere through the origin and tangent to the γ-plot at A is pierced by the γ-plot as shown in (a); (b) represents the limiting condition for stability of a surface with respect to forming facets of S. The sphere tangent at A passes through S.

A and S. The limiting conditions will be when the γ-plot point representing S lies on the sphere and the γ-plot and the sphere have a common tangent at S. These two conditions may be expressed by the equations for the stability of A:

$$\left(\frac{\partial \gamma}{\partial \theta}\right)_S \geqslant \gamma_A \sin \theta_A + \left(\frac{\partial \gamma}{\partial \theta}\right)_A \cos \theta_A,$$

$$\gamma_S \geqslant \gamma_A \cos \theta_A - \left(\frac{\partial \gamma}{\partial \theta}\right)_A \sin \theta_A. \tag{1.14}$$

The equality applies when A and S coexist, a situation which would be obtained by the break-up of a surface of orientation between A and S (Fig. 1.7). When such faceting is observed (Figs. 1.8 and 1.9),

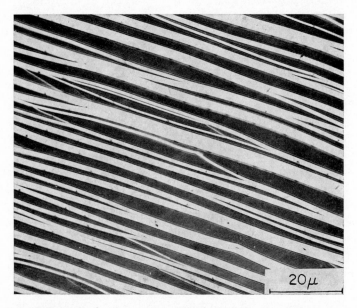

FIG. 1.8. Example of a silver crystal exhibiting linear faceting. The mean surface is inclined at about 10° to the (111) plane. The light bands of the optical micrograph are portions of (111) plane, while the dark regions are portions of "complex" surface. The crystal had been heated in air at 900°C for 10 days. (Courtesy A. J. W. Moore.)

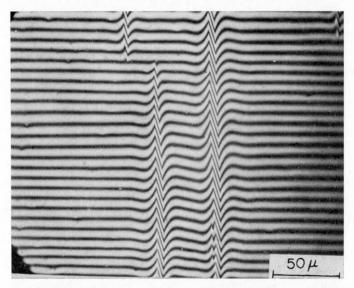

FIG. 1.9. Interferometric pattern from isolated facets on an otherwise flat surface of copper. (Courtesy of W. D. Robertson.)

eqns. (1.14) may be used to derive some information on the local shape of the γ-plot. For example, with the approximation that $(\partial\gamma/\partial\theta)_A$ is small we may obtain γ_A/γ_S and $1/\gamma_S(\partial\gamma/\partial\theta)_S$ from measurement of the angle of intersection S with A, i.e. the orientation of A.

Before leaving the question of surface stability (for the moment) it is appropriate to mention the influence of kinetic effects on the observations. It may be possible that irreversible evaporation of the crystal takes place at such a rate that the system is far from equilibrium, with the result that the surface configuration is determined not by free energy minimization but by kinetic parameters. Thus it has been argued that a number of examples of faceting are essentially unrelated to the Wulff plot and that the topography is due to the variation of evaporation rate with orientation. This point of view is well represented in the review article by Moore cited in the bibliography. Another interesting situation arises if the process of faceting is one which involves a nucleation barrier. Thus if faceting occurs by a continuous rotation of the surface normal of a portion of the surface, then an orientation such as A (Fig. 1.5) will be metastable. Alternatively, a nucleation barrier could arise simply from the extra free energy associated with edges of facet planes. The nonuniform nature of faceting and the occasional appearance of isolated facets on otherwise flat surfaces (Fig. 1.9) may be taken as evidence for a nucleation barrier to faceting.

1.5. INTERSECTION OF INTERFACES

Polycrystalline solids contain a variety of interfaces which include grain boundaries, stacking faults, and twin boundaries as well as free surfaces of various orientations. The problem of determining the overall arrangement of these interfaces which is the stable one is very complex, but the local equilibrium configuration at the intersection of interfaces is considerably simpler to predict and is also of some fundamental interest in connection with the determination of the relative surface tensions of the different interfaces. We consider three planar interfaces (Fig. 1.10) intersecting along a line through O and perpendicular to the plane of the paper and examine the region in the immediate vicinity of the intersection to obtain the configuration stable with respect to *small* displacements. Our derivation of the equilibrium configuration again follows the classic work of Herring. As we shall see later, a useful method for determining the γ-plot is based on these considerations.

Consider a small displacement of the line of intersection parallel to the boundary 1 so that boundary 1 is represented by bP, boundary 2 by aBP, and boundary 3 by cCP. Considering unit length perpendicular to the plane of the paper the change in surface tension or surface free energy is, to first order,

$$\delta F^s = \gamma_1(OP) + \gamma_2(BP - BO) + \gamma_3(CP - CO)$$

$$+ BP \frac{\partial \gamma_2}{\partial \alpha_2} \delta \alpha_2 + CP \frac{\partial \gamma_3}{\partial \alpha_3} \delta \alpha_3. \tag{1.15}$$

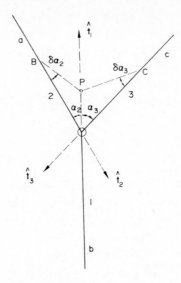

FIG. 1.10. Intersection of three planar boundaries along a line through O and perpendicular to the plane of the paper. To establish the equilibrium configuration we imagine the line of intersection to be displaced to P and examine the stability of the system with respect to such small displacements.

Evaluating the lengths in terms of the angles and using the condition for the initial configuration to be the equilibrium one, i.e. $\delta F^s = 0$, gives

$$\gamma_1 - \gamma_2 \cos \alpha_2 - \gamma_3 \cos \alpha_3 + \sin \alpha_2 \left(\frac{\partial \gamma_2}{\partial \alpha_2} \right) + \sin \alpha_3 \left(\frac{\partial \gamma_3}{\partial \alpha_3} \right) = 0. \quad (1.16)$$

If $\hat{t}_i$ is a unit vector lying in the plane of the diagram and in the plane of the boundary i as shown in Fig. 1.10, we may rewrite eqn. (1.16) as

$$\left(\gamma_1 \hat{t}_1 + \gamma_2 \hat{t}_2 + \gamma_3 \hat{t}_3 + \frac{\partial \gamma_1}{\partial \hat{t}_1} + \frac{\partial \gamma_2}{\partial \hat{t}_2} + \frac{\partial \gamma_3}{\partial \hat{t}_3} \right) \cdot \hat{t}_1 = 0. \quad (1.17)$$

Similarly, by considering displacements of the line of intersection along $\hat{t}_2$ and $\hat{t}_3$ we may generate two other equations analogous to (1.17). Since $\hat{t}_1$, $\hat{t}_2$, and $\hat{t}_3$ are nonparallel co-planar vectors the vector quantity in the square brackets must be identically zero. Thus the equilibrium configuration at the intersection of three planar interfaces may be written as

$$\sum_{i=1}^{3} \left(\gamma_i \hat{\mathbf{t}}_i + \frac{\partial \gamma_i}{\partial \hat{\mathbf{t}}_i} \right) = 0. \tag{1.18}$$

The vector quantities $\partial \gamma_i / \partial \hat{\mathbf{t}}_i$ have the mathematical form of a torque, i.e. the change in a free energy per unit of angular rotation, and are generally referred to in the literature as the *torque terms*. They may alternatively be viewed as a force per unit area acting normal to the interface.[†] The following are two special cases in which the torque terms are especially important in influencing interface morphologies.

(a) *Twin boundaries*

Twin boundaries are rather specialized internal interfaces in solids in that although the two crystals on either side of the boundary are related to each other by large orientation differences they have extremely small values of surface tension associated with them. For example, in copper the twin boundary tension is of the order of 30 ergs/cm² as compared to ~ 500 ergs/cm² for a normal high angle grain boundary. The reason for the low value of γ in this case is that one crystal can be obtained from the other by performing a reflection operation about the twin plane (of type {111} in face-centered cubic metals) and nearest neighbor coordination is not disturbed by the presence of the boundary. Small perturbations in the orientation of the boundary relative to the two crystals would thus be expected to lead to relatively large increases in tension. Thus we expect $\partial \gamma_i / \partial \hat{\mathbf{t}}_i$ to be very large for a twin and in such a direction as to maintain the boundary coincident with the twinning plane. This is the main reason that twin boundaries invariably produce straight traces on the surface of face-centered cubic metals (Fig. 1.11).

[†] Note that eqns. (1.18) are equivalent to (1.16) for the special case of the intersection of two interfaces.

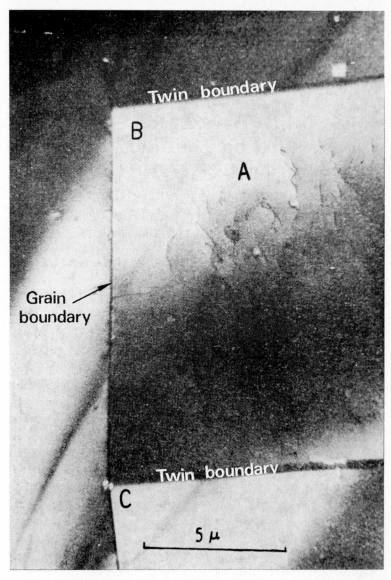

FIG. 1.11. Electron micrograph showing the intersection of a pair of twin boundaries with a grain boundary in a copper–1 wt.% antimony alloy. At C all three boundaries lie in the same 180°. (From M. C. Inman and H. R. Tipler, *Metallurgical Reviews* **8**, 105 (1963).)

(b) *Three intersecting boundaries all in the same 180°* (Figs. 1.11 and 1.12)

For fluid interfaces for which the quantities $\partial\gamma_i/\partial\hat{t}_i$ are zero it is obvious that the configuration at the intersection of three such bound-

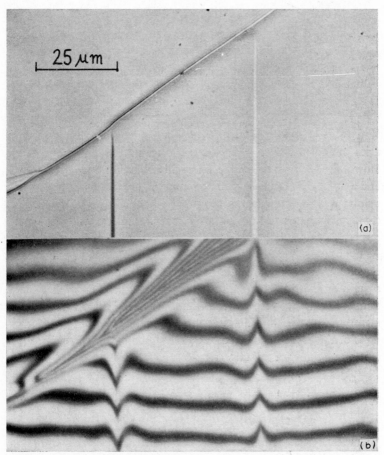

Fig. 1.12. (a) Intersection of a pair of twin boundaries with a free surface. (b) A two-beam interference microscope photograph showing the topography of the surface in (a). That all three boundaries can lie in the same 180° at their intersection is due to the influence of the variation of surface tension with surface orientation. (Courtesy of H. Mykura.)

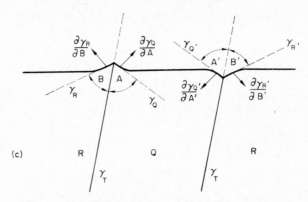

FIG. 1.12. (c) Definition of quantities used in eqn. (3.9).

aries cannot be such that they all lie in the same semicircle. However, for crystalline interfaces this situation becomes possible and a number of examples can be found in the literature. They usually involve a twin boundary or a low angle boundary which has a sufficiently small free energy such that the derivatives of the γ's with respect to orientation are significant in magnitude compared to the γ's themselves. Figures 11.1 and 1.12(a) and (b) show examples with three intersecting boundaries. In Fig. 1.11 a twin boundary intersects a grain boundary and in Fig. 1.12(b) we see the intersection of twin boundaries with a free surface. Later we shall see how quantitative measurements of the angles at the intersections and of the orientations of the crystals involved have been used to determine the Wulff plot of a number of crystals.

1.6. CURVED INTERFACES AND THE GIBBS–THOMPSON RELATION

Curved interfaces are involved in numerous practical situations with bubbles, droplets, and precipitate particles, and in a number of techniques for determining the surface tensions of both liquids and solids. There are two relationships which are of special interest in connection with curved surfaces. These are contained in the equation of Young and Laplace and in the Kelvin or Gibbs–Thompson relation. The first of these equations has to do with the excess pressure inside a spherical droplet and the second is concerned with the relationship between equilibrium vapor pressure and radius; they are not independent and, indeed, the second may be simply derived from the first. The Young and Laplace equation can be obtained for a liquid droplet by considering the condition of mechanical equilibrium in the presence of an isotropic surface stress g. At equilibrium the pressure inside the drop exceeds that outside by an amount ΔP given by

$$\Delta P = \frac{2g}{R}, \tag{1.19a}$$

$$= \frac{2\gamma}{R}, \tag{1.19b}$$

where R is the liquid drop radius. The second equation applies to an isotropic fluid for which surface stress and tension are identical. The increase in chemical potential caused by the pressure increase of eqn. (1.19) is to first order

$$\Delta \mu \simeq \Omega_0 \Delta P = \frac{2g}{R} \Omega_0, \tag{1.20}$$

where Ω_0 is the atomic volume. If we treat the vapor as an ideal gas for which the chemical potential is equal to $(kT \ln p + \text{constant})$ the vapor pressure p that is in equilibrium with the spherical particle is related to that over a flat surface p_0 by

$$kT \ln \left(\frac{p}{p_0}\right) = \Delta \mu \tag{1.21a}$$

or

$$kT \ln \left(\frac{p}{p_0}\right) = \frac{2\gamma}{R} \Omega_0. \tag{1.21b}$$

Equation (1.21b) is the usual form of the Gibbs–Thompson equation. This equation shows that the higher the curvature the greater the vapor pressure so that in a system containing a distribution of drop sizes the small ones should be expected to disappear by transfer to the larger particles, as is indeed observed.

For interfaces involving solids we might expect that relationships similar to eqns. (1.19) and (1.21) would exist. However, in such cases one has to be more careful about the distinction between surface stress and tension and also to recognize the orientational and directional dependence of these quantities. Thus the Young and Laplace equation should be replaced for a solid–vapor interface by a relationship between the stress distribution in the particle and the surface stress. However, the approximation that the surface stress is isotropic is often made in connection with solid surfaces. For example, for small crystalline particles the surface stress is assumed also to be given by eqn. (1.19a) which should give rise to an average fractional decrease in atomic volume given by

$$\frac{\Delta \Omega}{\Omega_0} = \beta \frac{2g}{R}$$

or for a cubic crystal, a fractional change in lattice parameter

$$\frac{\Delta a}{a} = \frac{1}{3}\beta\frac{2g}{R}, \tag{1.22}$$

where β is the isothermal bulk compressibility. A number of investigators have in fact reported deviations in the lattice parameters of small crystalline particles. We shall quote some of these measurements later but it is worth noting at this point that the resulting values of g obtained by using eqn. (1.22) are significantly different in magnitude in most cases from the corresponding surface tensions. Negative values of g corresponding to surfaces in compression have in fact been obtained in a few cases, whereas values of γ are invariably positive.

For interfaces between a solid and a liquid or a solid and a vapor, the Gibbs–Thompson relation is more complicated than eqn. (1.21) for the reasons noted above. However, for some purposes the directional effects are ignored and a reasonably satisfactory (if not quantitative) description of phenomena involving curved solid interfaces can often be given using the form of eqn. (1.21). A more general form of the Gibbs–Thompson equation has, however, been derived for the case where there is sufficiently high atomic mobility that the surface stress and tension may be identified but where the dependence of γ on orientation is included. We shall derive the form of the Gibbs–Thompson relationship, corresponding to eqn. (1.20), following arguments similar to those given by Herring.

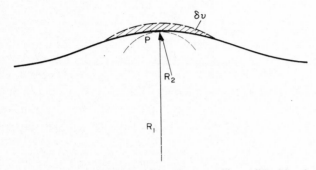

Fig. 1.13. Portion of a curved surface of a crystalline solid with principal radii of curvature R_1 and R_2. To examine the value of the chemical potential in the vicinity of any point P we imagine a small hump of volume δv to be created and apply the criterion for (local) equilibrium that the system should be stable against such perturbations. (After C. Herring, in *The Physics of Powder Metallurgy*, ed. W. E. Kingston, McGraw-Hill, 1951.)

Consider a portion of a surface having principal radii of curvature R_1 and R_2 (Fig. 1.13) and imagine the creation of a small hump of volume δv by the introduction of $\delta v / \Omega_0$ vacancies into the crystal at this portion. Ω_0 is the atomic volume. Then if μ_v is the local value of the vacancy chemical potential and p the mean hydrostatic pressure locally in the crystal, the change in the volume contribution to the Helmholtz free energy may be written as $-p\,\delta v + \mu_v(\delta v / \Omega_0)$. The increase in surface free energy is $\delta(\int \gamma\, dS)$. For equilibrium we require the net change in free energy to be zero, i.e.

$$\left(-p\,\delta v + \mu_v \frac{\delta v}{\Omega_0}\right) + \delta\left(\int \gamma\, dS\right) = 0$$

or

$$\mu_v = p\Omega_0 - \frac{\Omega_0}{\delta v}\, \delta\left(\int \gamma\, dS\right). \qquad (1.23)$$

The second term on the right hand side may be evaluated as shown by Herring in terms of the local radii of curvature, surface tension, and its derivatives with respect to orientation. We obtain

$$\mu_v = p\Omega_0 - \Omega_0\gamma\left\{\left(\frac{1}{R_1}+\frac{1}{R_2}\right)+\frac{1}{\gamma}\left(\frac{\partial^2\gamma}{\partial n_x^2}\frac{1}{R_1}+\frac{\partial^2\gamma}{\partial n_y^2}\frac{1}{R_2}\right)\right\}. \qquad (1.24)$$

The derivatives with respect to n_x and n_y measure changes in surface tension γ with changes in the orientation of the surface normal in the direction of the two principal curvatures. The chemical potential of atoms relative to its value μ_0 beneath a flat surface where the pressure is taken as zero is given by

$$\mu = \mu_0 + p\Omega_0 \qquad (1.25)$$

if we neglect small terms of order $x_v kT$, where x_v is the fractional vacancy concentration.

Combining eqns. (1.24) and (1.25),

$$\mu - \mu_v = \mu_0 + \Omega_0\gamma\left\{\left(\frac{1}{R_1}+\frac{1}{R_2}\right)+\frac{1}{\gamma}\left(\frac{\partial^2\gamma}{\partial n_x^2}\frac{1}{R_1}+\frac{\partial^2\gamma}{\partial n_y^2}\frac{1}{R_2}\right)\right\}. \qquad (1.26)$$

If vacancies are always in local equilibrium so that we may set $\mu_v = 0$ everywhere we obtain, corresponding to eqn. (1.20),

$$\Delta\mu = \Omega_0\gamma\left\{\left(\frac{1}{R_1}+\frac{1}{R_2}\right)+\frac{1}{\gamma}\left(\frac{\partial^2\gamma}{\partial n_x^2}\frac{1}{R_1}+\frac{\partial^2\gamma}{\partial n_y^2}\frac{1}{R_2}\right)\right\}. \qquad (1.27)$$

It is clear that this generalized Gibbs–Thompson relationship reduces to that of a liquid when γ may be regarded as isotropic. We shall use eqn. (1.27) in discussing mass transport near the surfaces of crystals in section 7.3. For the moment we note that gradients in chemical potential can be generated by gradients in surface curvature and that these will produce atomic fluxes. The magnitudes of the deviations in chemical potential are quite small for all but the smallest radii of curvature. For example, for a radius of curvature of 1 micron (10^{-6} m), $\Delta\mu/kT \approx 10^{-3}$ at 1000°K for a typical metal. The measurement of the rate at which these differences in chemical potential are eliminated in crystals provides information on parameters such as volume or surface diffusivities.

So far in our discussion of one-component systems we have not considered the question of the equilibrium structure of the interface. This question cannot, of course, be answered completely for any particular case within the framework of thermodynamics since a specific model of atomic interactions must be adopted at some point in the development. However, the general framework for the description of the transition or inhomogeneous region between two phases and the general concept of diffuse interfaces are of considerable interest. We shall take up this topic in Chapter 2 in connection with interfaces in multicomponent systems. For the moment we may note that an interface between two phases should not be regarded as a mathematical plane at which there is a discontinuous change in properties from those characteristic of one bulk phase to those characteristic of the other bulk phase. Rather the transition should be regarded as occurring over a region of finite width (as indicated schematically in Fig. 1.1). This effect will be most marked for liquid–vapor interfaces or liquid–liquid interfaces near critical points where the interface essentially becomes infinitely diffuse as the two phases become identical.

In dealing with curved interfaces in section 1.6 we have implicitly assumed that the value of the surface tension was independent of the local radius of curvature of the surface. There are actually a number of published papers dealing with the question of the dependence of γ on curvature for liquid systems. It appears to be generally agreed that γ will not depart from its value characteristic of a macroscopic flat surface until the particle radius becomes comparable to atomic dimensions. This is intuitively reasonable since deviations from the macroscopic value should be significant only when the curvature is sufficiently large that the surface coordination is changed or that there is a distortion of the transition region. The exact way in which γ varies with curvature will depend on the type of atomic interactions that are involved. The usefulness of macroscopic thermodynamic quantities such as surface tension is questionable in any event when the system contains only a small number of atoms.

BIBLIOGRAPHY

ADAMSON, A. W., Physical Chemistry of Surfaces, 2nd ed., Interscience, New York, 1967.

CABRERA, N., The equilibrium of crystal surfaces, Surface Science 2, 320 (1964).

CABRERA, N., and COLEMAN, R. V., Theory of crystal growth from the vapor, in The Art and Science of Growing Crystals (ed. J. J. Gilman), Wiley, New York, 1963.

FLOOD, E. A. (ed.), The Gas–Solid Interface, Vol. I, Dekker, New York, 1967.

GIBBS, J. W., The Collected Works of J. Willard Gibbs, Longmans, New York, 1928.

GRUBER, E. E., and MULLINS, W. W., On the theory of anisotropy of crystalline surface tension, J. Phys. Chem. Solids 28, 875 (1967).

GUGGENHEIM, E. A., Thermodynamics, 4th edn., North-Holland, Amsterdam, 1959.

HERRING, C., (a) The use of classical macroscopic concepts in surface energy problems, in Structure and Properties of Solid Surfaces (ed. R. Gomer and C. S. Smith), Univ. of Chicago Press, 1953, Chap. 1. (b) Surface tension as a motivation for sintering, in The Physics of Powder Metallurgy (ed. W. E. Kingston), McGraw-Hill, New York, 1951.

HIRTH, J. P., The kinetic and thermodynamic properties of surfaces, in Energetics in Metallurgical Phenomena (ed. W. M. Mueller), Gordon & Breach, 1965.

INMAN, M. C., and TIPLER, H. R., Interfacial energy and composition in metals and alloys, Metallurgical Reviews 8, 105 (1963).

MOORE, A. J. W., Thermal faceting, in Metal Surfaces, American Society for Metals, Cleveland, 1963.

MULLINS, W. W., Solid surface morphologies governed by capillarity, in Metal Surfaces, American Society for Metals, Cleveland, 1963.

WINTERBOTTOM, W. K., Crystallographic anisotropy in the surface energy of solids, in Surfaces and Interfaces I (ed. J. J. Burke, N. L. Reed, and V. Weiss), Syracuse University Press, 1967.

CHAPTER 2

MULTICOMPONENT
SYSTEMS

2.1. DEFINITIONS

Although there are a number of interesting properties associated with interfaces in one-component systems, many of the important surface phenomena involve two or more components. Adsorption is a necessary step in many surface reactions. This includes the binding to the surface of atoms or molecules from the vapor phase and the segregation of impurities from the crystal interior to the surface. A number of challenging problems concerned with the binding and structure of adsorbed layers remain to be solved. The thermodynamic description of adsorption is relatively straightforward, and we shall cover some of the main features in the present chapter.

Consider an interface between two phases in a multicomponent system. Figure 2.1 shows possible concentration profiles normal to a planar interface in a binary system. We may define surface excess quantities by introducing a dividing surface exactly as in the one-component case so that the surface excess Γ_i of any component i per unit area of interface is given by $\Gamma_i = n_i^s s/A$, where A is the interface area and

$$n_i^s = n_i^{\text{total}} - (c_i^\alpha V^\alpha + c_i^\beta V^\beta), \tag{2.1}$$

where n_i^{total} is the total number of moles of type i, c_i^α and c_i^β are the molar concentrations of i characteristic of the bulk α and β phases and V^α and V^β are the volumes that the α and β phases would have if they continued uniform to the dividing surface. The surface tension

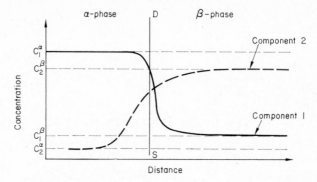

Fig. 2.1. Possible variation with distance of the concentrations of the two components in a binary system across the interface region between two phases α and β. DS denotes a dividing surface with respect to which excess quantities are defined.

may also be defined as in the one-component case as the amount of work involved in creating unit area of new surface at constant temperature, volume, and total number of moles of each component. Following arguments similar to those involved in deriving eqn. (1.5) we obtain for the relationship between surface tension, specific surface free energy, and excesses

$$\gamma = f^s - \sum_i \Gamma_i \mu_i, \qquad (2.2)$$

where

$$\mu_i = \left(\frac{\partial F^{\text{total}}}{\partial n_i} \right)_{T, v, n_{j \neq i}}.$$

Except in very special cases the second term on the right hand side will not vanish for systems of two or more components, so that the surface tension and specific surface free energy are not in general equal. The distinction will become more apparent when we consider the framework for evaluating these quantities in terms of a particular solution model.

As in the one-component case, the quantities surface stress and surface tension will be numerically equal for liquids, whereas for multicomponent solids there will in general be three distinct and different quantities: surface tension, surface stress, and specific surface free energy.

2.2. ADSORPTION ISOTHERMS

Adsorption isotherms are graphs of or expressions for the surface excesses of the various components at constant temperature in terms of the thermodynamic properties of the mixture and the surface properties such as surface tension.

Consider a system which is closed in the sense that the total number of molecules of any particular type remains fixed. To find the relationship between surface excess and other thermodynamic properties we consider a small perturbation of the system at equilibrium. For any such perturbation the change in the Helmholtz free energy of the entire system is given by

$$dF^{\text{total}} = -S^{\text{total}} \, dT - P \, dV^{\text{total}}$$

or

$$-S^{\text{total}} \, dT - P \, dV^{\text{total}} = dF^\alpha + dF^\beta + A \, df^s$$
$$= dF^\alpha + dF^\beta + A \, d\gamma + A \sum_i \Gamma_i \, d\mu_i + A \sum_i \mu_i \, d\Gamma_i$$

$$(2.3)$$

from eqn. (2.2). Although we may maintain the entire system closed in any perturbation, the material may be redistributed between the two phases and there may also be changes in surface excesses. Thus

$$dF^\alpha = -S^\alpha \, dT - P \, dV^\alpha + \sum_i \mu_i \, dn_i^\alpha, \qquad (2.4)$$

and there will, of course, be a similar equation for the β-phase. Combining with eqn. (2.3) then gives

$$d\gamma = -s^s \, dT - \sum_i \Gamma_i \, d\mu_i, \qquad (2.5)$$

where s^s is the specific surface entropy. Setting $dT = 0$ we obtain the famous *Gibbs adsorption isotherm*

$$dy = -\sum_i \Gamma_i \, d\mu_i$$

or

$$\Gamma_i = -\left(\frac{\partial \gamma}{\partial \mu_i}\right)_{T, n_j, \mu_{j \neq i}},$$

(2.6)

relating the surface excesses to the variation in surface tension with chemical potentials of the components.

The term adsorption is generally taken to mean the accumulation at a solid–gas or liquid–gas surface of atoms or molecules which are normally present in the gas phase. It is clear from the derivation that the isotherm (2.6) also applies to the phenomenon of segregation in which one or more components from the interior of a solid or liquid become concentrated or depleted in the immediate vicinity of an interface. As will be evident from the examples given below, a particular model for the chemical potentials or the activities of the components will have to be assumed in order to relate γ and the Γ_i to the concentrations in the bulk solution.

(a) *Solid–gas interface in a two-component system*

Possible concentration profiles for such a system are shown schematically in Fig. 2.2, where some solubility of the gas atoms (component 2) in the solid is indicated and the solid has a small vapor pressure. In this case it will be possible to locate the dividing surface such that $\Gamma_1 = 0$.

Thus

$$dy = -\Gamma_2 \, d\mu_2 \quad \text{at constant temperature}$$

or

$$\Gamma_2 = -\left(\frac{\partial \gamma}{\partial \mu_2}\right)_{T, n_1, n_2}.$$

(2.7)

Assuming ideal gas behavior, which is usually a very good approximation except at very high pressures ($\gtrsim 10$ atm),

$$\mu_2 = \mu_2^0 + RT \ln p_2,$$

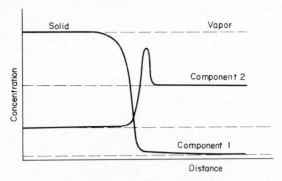

FIG. 2.2. Possible concentration profiles across a solid–gas interface. Some solubility of the gas atoms (component 2) in the solid phase is indicated as well as an accumulation of gas in the interface region.

where μ_2^0 is the chemical potential in a standard state of gas at one atmosphere and temperature T. Hence the Gibbs adsorption isotherm for this particular case becomes

$$\Gamma_2 = -\frac{1}{RT}\left(\frac{\partial\gamma}{\partial\ln p_2}\right)_T, \tag{2.8}$$

which states the fact, perhaps intuitively obvious, that gases will tend to accumulate on surfaces when increase in their pressure causes a decrease in surface tension. There are a number of measurements of the variation of γ with gas pressure for liquid–gas interfaces. The data on solid–gas interfaces are, however, relatively scarce since solid surface tensions are much more difficult to measure. Figure 2.3 illustrates some measurements of the average surface tension of polycrystalline silver with variations in the pressure of oxygen. The slope in the high pressure region is constant and corresponds to a surface excess of oxygen of about 10^{15} atoms/cm^2, i.e. a constant coverage of the surface under these temperature and pressure conditions of about one monolayer. One would expect a levelling off at sufficiently low pressures to a constant value of γ as the surface excess approaches zero. Considerable care has to be exercised in using eqn. (2.8) since it must

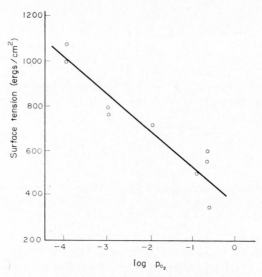

FIG. 2.3. Variation of (average) surface tension of silver with oxygen partial pressure. The slope of the line indicates an approximately constant excess of oxygen [from eqn. (2.8)] of about one monolayer over most of the pressure range studied. There is some indication of a decrease in the amount of adsorbed oxygen at the lower pressures. After F. H. Buttner, E. R. Funk, and H. Udin, *J. Phys. Chem.* **56**, 657 (1952).

be remembered that it applies only to adsorption and desorption processes which are reversible with changes in pressure. When the adsorbed gas molecules are very strongly bound to the surface the rate of desorption may be so slow that the equilibrium coverage will not be reached in the time of normal experimental observations.

(b) *Binary solid in contact with its own vapors*

Again it will usually be possible to choose the dividing surface such that the excess of the major constituent, say, is zero, and hence eqn. (2.7) will apply to the solute. An example of interest is that of a crystal containing a solute in a sufficiently small concentration that it is regarded as an impurity. The chemical potential of the impurity may then be expressed in the dilute solution approximation (which neglects all impurity–impurity interactions) as

$$\mu_2 = \mu_2^0 + RT \ln \gamma' x_2,$$

where γ' is the activity coefficient of the impurity, x_2 its mole fraction in the bulk, and γ' is independent of x_2. Hence, the surface excess is related to the impurity content by

$$\Gamma_2 = -\frac{1}{RT}\left(\frac{\partial \gamma}{\partial \ln x_2}\right)_T, \qquad (2.9)$$

which is essentially identical to (2.8). Figure 2.4 shows some experimental results on the variation of the surface tension of solid copper with mole fraction of dissolved antimony for a concentration range where the dilute solution approximation should hold. The antimony

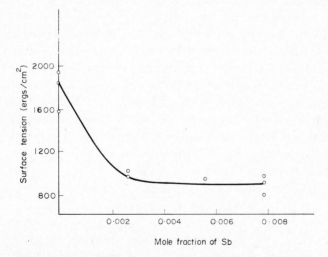

Fig. 2.4. Variation of the average value of the surface tension of copper with concentration of dissolved antimony. The rapid initial decrease in γ with increasing Sb content followed by the region of almost constant surface tension indicates that the Sb segregates to the surface and nearly saturates the surface with fairly small bulk Sb concentration. (From E. D. Hondros and D. McLean in *Surface Phenomena of Metals*, Society of Chemical Industry, Monograph No. 28. London, 1968, p. 39.)

apparently segregates to the surface, causing a rapid decrease in surface tension with increasing antimony concentration. Experimental support for the validity of the Gibbs adsorption isotherm has been obtained in this particular case by comparing the antimony excess predicted by eqn. (2.9) with that determined by direct measurements using a radioactive tracer technique.

The phenomenon of solute segregation to interfaces, particularly grain boundaries in alloys, is of considerable importance in influencing the mechanical properties. It is at present being studied by a wide variety of experimental methods, notably field ion microscopy, Auger electron spectroscopy, and ion bombardment techniques (Chapter 6).

2.3. STATISTICAL THERMODYNAMICS OF SURFACES IN BINARY SYSTEMS

The literature on the thermodynamics and statistical mechanics of interfaces in binary systems is very extensive. In the present section we will consider only some of the simple fundamental aspects of the statistical treatment and show how the concentration profile near a free surface and the surface tension for a binary crystal can be derived on the basis of a simplified model of the solution. The simplest models which have been used extensively in the discussion of the properties of solutions are the ideal solution and regular solution models, and we shall adopt these for our present purpose. The ideal solution may be defined by

$$\Delta \bar{H}_{\text{mix}} = 0,$$
$$\Delta \bar{S}_{\text{mix}} = -R(x_1 \ln x_1 + x_2 \ln x_2), \tag{2.10}$$

where $\Delta \bar{H}_{\text{mix}}$ and $\Delta \bar{S}_{\text{mix}}$ are respectively the heat and entropy of mixing per mole of solution. In terms of a model involving only nearest neighbor pair interactions, the criterion for an ideal solution is

$$\varepsilon_{12} = \tfrac{1}{2}(\varepsilon_{11} + \varepsilon_{22}), \tag{2.11}$$

where ε_{11} is the interaction energy between a pair of atoms of type 1, etc. In the regular solution the molar entropy of mixing is given by eqns. (2.10), but the heat of mixing is expressed in the form

$$\Delta \overline{H}_{mix} = \omega\, x_1 x_2 N_0, \tag{2.12}$$

where N_0 is Avogadro's number and ω a constant. In terms of a model of random substitutional solid solution with nearest neighbor pair interactions only, the regular solution requires

$$\omega = \frac{z}{2}\, \Delta \varepsilon$$

with

$$\Delta \varepsilon = 2\varepsilon_{12} - (\varepsilon_{11} + \varepsilon_{22}), \tag{2.13}$$

where z is the coordination of an atom in the solid.[†]

From eqn. (2.2) the *total* surface tension associated with a planar interface between two phases α and β is given by

$$A\gamma = F^s - \sum_i n_i^s \mu_i,$$

and since

$$\sum_i n_i^s \mu_i = \sum_i n_i^{\text{total}} \mu_i - \sum_i (n_i^\alpha \mu_i + n_i^\beta \mu_i),$$

$$A\gamma = (F^{\text{total}} - G^{\text{total}}) - \{(F^\alpha - G^\alpha) + (F^\beta - G^\beta)\}$$

$$= \Omega^{\text{total}} + (PV^\alpha + PV^\beta),$$

i.e.

$$A\gamma = \Omega^{\text{total}} + PV, \tag{2.14}$$

recalling that for a homogeneous phase $(F - G) = -PV$. Now if Z is the grand partition function[‡] of the real system,

$$\Omega^{\text{total}} = -kT \ln Z,$$

[†] The parameter $\Delta \varepsilon$ is related to the critical temperature T_c above which the miscibility gap disappears by $T_c = (z\Delta\varepsilon)/2k$. (See E. A. Guggenheim, *Thermodynamics*, North-Holland, Amsterdam, 1959, p. 256.)

[‡] See, for example, T. Hill, *Introduction to Statistical Thermodynamics*, Addison-Wesley, 1960.

and hence

$$A\gamma = -kT \ln Z + PV. \qquad (2.15)$$

Alternatively, if Z^α is the grand partition function of a homogeneous volume V^α of phase α (i.e. the volume of the α-phase if it continued homogeneous to the dividing surface) at fixed temperature T and chemical potentials μ_i, and Z^β the corresponding quantity for the β-phase,

$$A\gamma = -kT \ln \frac{Z}{Z^\alpha Z^\beta}. \qquad (2.16)$$

The total Helmholtz free energy F is related to the canonical partition function Q by

$$F = -kT \ln Q,$$

so that it follows that the specific surface free energy f^s may be expressed as

$$A f^s = -kT \ln \frac{Q}{Q^\alpha Q^\beta}, \qquad (2.17)$$

where Q^α and Q^β are the canonical partition functions of the two phases assuming that they remained homogeneous to the dividing surface.

Equations (2.16) and (2.17) form the link between the macroscopic quantities and the partition functions which are in turn connected to the microscopic model by the definitions of Q and Z, i.e.

$$Q = Q(N_1, N_2, V, T) = \sum_{\text{states } j} \exp\left(-\frac{E_j(N_1, N_2, V)}{kT}\right) \qquad (2.18)$$

where $E_j(N_1, N_2, V)$ is the energy of the system of volume V containing N_1 molecules of type 1 and N_2 molecules of type 2 when in a particular state j and

$$Z = \sum_{N_1, N_2, j} \exp\left(\frac{N_1 \mu_1}{kT}\right) \exp\left(\frac{N_2 \mu_2}{kT}\right) \exp\left(-\frac{E_j(N_1, N_2, V)}{kT}\right). \qquad (2.19)$$

2.3.1. *Free Surface of a Binary Solid Solution*

As mentioned above, we will treat the solution as one in which the atoms interact as nearest neighbor pairs only; although somewhat unrealistic, this model nevertheless serves to demonstrate many of the important features of the problem. The present development follows the early work of Ono and others (see bibliography) applied to the free surface of a crystalline solution.

We suppose that the crystal consists of a set of planes parallel to the free surface and that $x_1^{(r)}$ and $x_2^{(r)}$ are the atomic fractions of components 1 and 2 in the rth plane below the surface (Fig. 2.5).

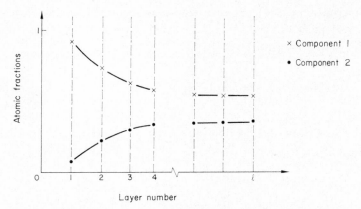

FIG. 2.5. Schematic representation of a possible variation in composition from one plane to the next near the free surface of a binary alloy.

(We assume that the density of atoms above plane 1, i.e. in the vapor, is sufficiently small that interactions with these atoms may be neglected.) We consider a planar interface of area A with n sites per unit area on each plane with $x_1^{(l)}$ and $x_2^{(l)}$ the atomic fractions deep in the crystal.

Then

$$x_1^{(r)} + x_2^{(r)} = 1 \tag{2.20}$$

and the total numbers of atoms of the two types in a crystal consisting of l lattice planes are

$$N_1 = nA \sum_{r=1}^{l} x_1^{(r)}; \qquad N_2 = nA \sum_{r=1}^{l} x_2^{(r)}. \qquad (2.21)$$

In the interest of minimizing the number of symbols we shall suppose that we are dealing with a (111) surface of a face-centered cubic crystal so that the number of nearest neighbors of a particular atom in the same plane is 6 and is 3 in each of the adjacent planes (see section 4.2, Fig. 4.1). We shall assume that within each layer there is complete randomness in the distribution of the atoms of types 1 and 2 (the Bragg–Williams approximation). To obtain the energy of any particular arrangement, i.e. $E(x_1^{(r)}, x_2^{(r)})$, we require to obtain the number of neighboring pairs of types 1–1, 2–2, and 1–2 respectively. It is readily shown that these are given by

$$\text{Number of } 1\text{–}1 \text{ bonds} = 3nA \sum_{r=1}^{l} x_1^{(r)} X_1^{(r)} = M_{11},$$

$$\text{Number of } 2\text{–}2 \text{ bonds} = 3nA \sum_{r=1}^{l} x_2^{(r)} X_2^{(r)} = M_{22}, \qquad (2.22)$$

$$\text{Number of } 1\text{–}2 \text{ bonds} = 3nA \sum_{r=1}^{l} [x_1^{(r)} X_2^{(r)} + x_2^{(r)} X_1^{(r)}] = M_{12},$$

where

$$X_i^{(r)} = x_i^{(r)} + \tfrac{1}{2}(x_i^{(r+1)} + x_i^{(r-1)}).$$

The energies to be used in the grand partition function (eqn. (2.19)) are then given by

$$E(x_1^{(r)}, x_2^{(r)}) = (M_{11}\varepsilon_{11} + M_{22}\varepsilon_{22} + M_{12}\varepsilon_{12}). \qquad (2.23)$$

The most probable concentration profile is the one which corresponds to the maximum term of Z and can be obtained by requiring that the chemical potentials for the two species are uniform throughout the solution; after considerable manipulation this condition yields the set of equations

$$(\mu_1-\mu_2)+6(\varepsilon_{11}-\varepsilon_{22}) = \frac{3}{2}(\varepsilon_{11}-\varepsilon_{22})+3\left\{(x_2^{(1)}-x_1^{(1)})+\frac{1}{2}(x_2^{(2)}-x_1^{(2)})\right\}$$

$$\times \Delta\varepsilon + kT\ln\frac{x_1^{(1)}}{x_2^{(1)}}$$

$$= 3(X_2^{(r)}-X_1^{(r)})\,\Delta\varepsilon+kT\ln\frac{x_1^{(r)}}{x_2^{(r)}} \qquad (2.24)$$

$$\text{for} \quad r = 2, 3, \ldots l,$$

whose solution for the $x_1^{(r)}$ defines the concentration profile near the interface. For this equilibrium concentration profile the grand partition function Z can then be shown to be given by

$$-kT\ln Z = nA\sum_{r=1}^{l}(\phi_1^{(r)}-x_1^{(r)}\mu_1-x_2^{(r)}\mu_2) \qquad (2.25)$$

with

$$\phi^{(1)} = -6[x_1^{(1)}\varepsilon_{11}+x_2^{(1)}\varepsilon_{22}-\{\tfrac{1}{2}x_1^{(1)}x_2^{(1)}+\tfrac{1}{8}(x_1^{(1)}x_2^{(2)}+x_1^{(2)}x_2^{(1)})\}\,\Delta\varepsilon$$
$$-\tfrac{1}{4}(x_1^{(1)}\varepsilon_{11}+x_2^{(1)}\varepsilon_{22})]+kT(x_1^{(1)}\ln x_1^{(1)}+x_2^{(1)}\ln x_2^{(1)})$$

and

$$\phi^{(r)} = -6\{x_1^{(r)}\varepsilon_{11}+x_2^{(r)}\varepsilon_{22}-\tfrac{1}{2}(x_1^{(r)}X_2^{(r)}+x_2^{(r)}X_1^{(r)})\,\Delta\varepsilon\}$$
$$+kT(x_1^{(r)}\ln x_1^{(r)}+x_2^{(r)}\ln x_2^{(r)}) \quad \text{for} \quad r = 2, 3, \ldots l. \ (2.26)$$

To determine the surface tension we require the grand partition function (Z^α in eqn. (2.16)) for a system consisting of l equivalent layers each of area A and with molecular fractions $x_1^{(l)}$ and $x_2^{(l)}$. If we use eqn. (2.25) for a set of l equivalent layers it is seen that this is given by

$$-kT\ln Z_1^\alpha = nAl(\phi^{(l)}-x_1^{(l)}\mu_1-x_2^{(l)}\mu_2) \qquad (2.27)$$

with

$$\phi^{(l)} = -6(x_1^{(l)}\varepsilon_{11}+x_2^{(l)}\varepsilon_{22}-x_1^{(l)}x_2^{(l)}\,\Delta\varepsilon)+kT(x_1^{(l)}\ln x_1^{(l)}+x_2^{(l)}\ln x_2^{(l)}).$$

Substituting eqns. (2.25), (2.26) and (2.27) into eqn.(2.16) and setting the grand partition function Z^β for the vapor phase equal to unity in the low pressure limit gives, finally,

$$A\gamma = An \sum_{r=1}^{l} (\phi^{(r)} - \phi^{(l)}) - n\mu_1 \sum_{r=1}^{l} (x_1^{(r)} - x_1^{(l)}) - n\mu_2 \sum_{r=1}^{l} (x_2^{(r)} - x_2^{(l)})$$

or

$$\gamma = n \sum_{r=1}^{l} (\phi^{(r)} - \phi^{(l)}) - \mu_1 \Gamma_1 - \mu_2 \Gamma_2, \tag{2.28}$$

where Γ_1 and Γ_2 are, as before, the surface excesses of the two components. The solution of the set of eqns. (2.24) provides all the unknowns in eqn. (2.28), so that once the equilibrium concentration profile is determined the surface tension can be computed. The parameters which we have to supply to determine the distribution at any particular temperature are ε_{11}, ε_{22}, and $\Delta\varepsilon$. In Fig. 2.6 are shown the results of some numerical computations for a particular set of these parameters. Computations of this type, using a nearest neighbor model, will generally lead to the result that the concentrations are significantly different only in the top one or two layers. With a more realistic interaction potential the mathematical problem becomes considerably more complex, and we should expect longer range deviations in concentrations in some cases. As yet there have been few quantitative experimental studies of the distribution of the components near the free surfaces of alloys. Some preliminary results from recent work on the problem using the techniques of Auger electron spectroscopy (section 6.4.1) and ion sputtering indicate that the concentrations of impurities in metallic crystals approach the average bulk values within a few atomic layers. The effect can nevertheless be of considerable importance, particularly with the surface of catalysts where small amounts of impurities can have a drastic effect on the catalytic efficiency.

Using eqn. (2.24) it is straightforward to show that the surface tension given by eqn. (2.28) for this particular model obeys the Gibbs adsorption isotherm, i.e.

$$d\gamma = -\Gamma_1 \, d\mu_1 - \Gamma_2 \, d\mu_2.$$

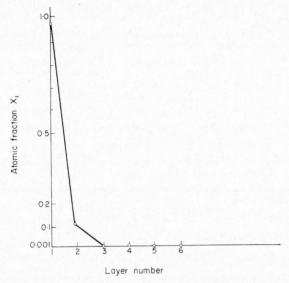

FIG. 2.6. Example of a calculated concentration profile near the surface of
a regular solution in which there are nearest neighbor interactions only.
The results are for a particular choice of the energy parameters. The calcu-
lations refer to a temperature T such that $T/T_e \approx 1$, where $T_e = z/2 \cdot \Delta\varepsilon/k$.
(From results listed by S. Ono, *Memoirs of the Faculty of Engineering, Kyu-
shu Univ.* **12**, 1 (1950).) The bulk atom fraction of component 1 is taken as
0·001. It is essentially reached after three layers in this particular case. This
is due to the short range of the assumed interaction.

Alternatively, we may regard this as a derivation of the isotherm for
the regular solution model. Since $\Gamma_1 = -\Gamma_2$ in this particular model,
some further manipulation using eqn. (2.24) leads to

$$\Gamma_2 = -x_1^{(l)}x_2^{(l)}(kT - 2x_1^{(l)}x_2^{(l)}\,\Delta\varepsilon)^{-1}\left(\frac{\partial\gamma}{\partial x_2^{(l)}}\right)_T, \qquad (2.29)$$

the form of the Gibbs adsorption isotherm for a regular solution,
first derived by Hildebrand.

There are two limiting cases of the model used here which are of
some interest and which also help to illustrate the limitations of the
present approach.

(a) *Pure solid*

If we do not allow for the possibility of lattice defects, then we obtain $x_1^{(r)} = 1$ for all r, and from eqns. (2.26) and (2.28)

$$\gamma = \tfrac{3}{2}n\varepsilon_{11}, \tag{2.30}$$

which is the answer obtained simply by counting the number of broken bonds at the surface, a procedure which generally greatly overestimates the magnitude of the surface tension (see section 5.3).

(b) *Ideal solution*

Setting $\Delta\varepsilon = 0$ leads to $\Gamma_1 = \Gamma_2 = 0$, i.e. uniform composition to the outermost plane, and from eqn. (2.28)

$$\gamma = x_1\gamma_1 + x_2\gamma_2, \tag{2.31}$$

where γ_1 and γ_2 are the surface tensions of the pure solids as given by eqn. (2.30). An approximately linear variation of γ with atomic fraction has in fact been observed in a number of liquid systems, and an example is shown in Fig. 2.7(a) for alloys of copper and nickel which is a system in which the derivations from ideality are quite small as established from thermodynamic activity measurements. Most other systems are considerably more complex and frequently exhibit maxima or minima in the variation of surface tension with composition. Data on the surface tension of solid alloys are very sparse due to the difficulty of the experimental measurement. Figure 2.7(b) shows some results on the variation with composition of the surface tension of solid Cu–Au alloys; this is an alloy system which shows considerable deviation from ideality in the solid state and in particular exhibits order–disorder transformations around the composition ratios 3:1, 1:1, and 1:3. The surface tension minimum is close to the composition Cu–Au which is the composition where a number of other properties including the electrical and thermal conductivity have extrema.

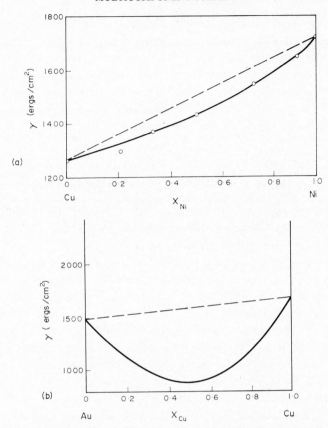

FIG. 2.7. (a) Variation of surface tension of Cu–Ni liquid alloys at 1550°C. Equation (2.31) is approximately obeyed in this system. (From J. G. Eberhart, *J. Phys. Chem.* **70,** 1183 (1966).) (b) Corresponding plot for solid alloys of gold and copper at 1000°C. (From results of S. S. White, C. M. Adams, and J. Wulff.)

The discussion in this section of the surfaces of binary alloys is intended to demonstrate the connection between the macroscopically observable surface tension and the microscopic model used to represent the interactions in the solution. The nearest neighbor quasichemical model allows the problem to be tractable but still complicated. With more realistic interatomic interactions the complexity of the problem increases tremendously, and it becomes difficult even to predict qualitative trends. An alternative approach for the description of transition regions is one in which the concentrations are considered as continuous functions of position. Such an approach was suggested in early work of van der Waals and has recently been developed by Cahn and Hilliard. It is clear that it will be most appropriate for cases with slowly varying composition.

2.4. CONTINUUM APPROACH TO THE DESCRIPTION OF INTERFACES

The basic postulate of this theory is that the free energy *per atom or molecule f* in a region of nonuniform composition depends not only on the value of the local composition but also on the local values of the derivatives of composition. (This is, of course, taken into account in the discrete layer model with nearest neighbor interactions where the binding energy of a molecule in the rth layer depends not only on the $x_i^{(r)}$ but also on the values of $x_i^{(r+1)}$ and $x_i^{(r-1)}$.) The treatment of f as a continuous function appears to limit the applicability of this method as noted above to situations involving only small gradients, such as may exist for liquid–vapor interfaces near critical points, and the approach will probably not be realistic for solid or liquid solution surfaces away from critical points.

We will assume for the sake of illustration that we have a planar interface with gradients in composition normal to the interface only. We will omit here lattice defects and consider a planar interface in a binary system separating two phases α and β as in the discussion of the previous section. Asserting that the free energy per molecule of the solution depends on the derivatives of the composition leads to the following expansion:

$$f\left(x_2, \frac{\partial x_2}{\partial z}, \frac{\partial^2 x_2}{\partial z^2}, \ldots\right) = f(x_2, 0, 0, \ldots) + L\left(\frac{\partial x_2}{\partial z}\right) + K_1\left(\frac{\partial^2 x_2}{\partial z^2}\right)$$
$$+ K_2\left(\frac{\partial x_2}{\partial z}\right)^2 + \ldots, \tag{2.32}$$

where x_2 is the atomic fraction of component 2 and z is now the co-ordinate normal to the interface. The coefficients are given by

$$L = \frac{\partial f}{\partial\left(\dfrac{\partial x_2}{\partial z}\right)}; \qquad K_1 = \frac{\partial f}{\partial\left(\dfrac{\partial^2 x_2}{\partial z^2}\right)}; \qquad K_2 = \frac{1}{2}\,\frac{\partial f}{\partial\left(\dfrac{\partial x_2}{\partial z}\right)^2},$$

$$(2.33)$$

and all derivatives are to be evaluated at the particular composition denoted by x_2. The total free energy of a volume V of the solution may then be written as

$$F = \int_v N_v\left\{f(x_2, 0, 0, \ldots) + K_1\frac{\partial^2 x_2}{\partial z^2} + K_2\left(\frac{\partial x_2}{\partial z}\right)^2 + \ldots\right\} dV,$$

$$(2.34)$$

where N_v is the number of molecules per unit volume. Following Cahn and Hilliard we have assumed $L = 0$ for a centrosymmetric crystal as otherwise the free energy per molecule would depend on the sign of the gradient in composition within a phase.

For an area A of a planar interface separating two phases (2.34) gives

$$F = A\int N_v\left\{f(x_2, 0, 0, \ldots) + K_1\left(\frac{\partial^2 x_2}{\partial z^2}\right) + K_2\left(\frac{\partial x_2}{\partial z}\right)^2 + \ldots\right\} dz,$$

$$(2.35)$$

and using Green's theorem to evaluate the second term on the right hand side,

$$F = A\int N_v\left\{f(x_2, 0, 0, \ldots) + K\left(\frac{\partial x_2}{\partial z}\right)^2 + \ldots\right\} dz, \qquad (2.36)$$

where

$$K = -\frac{dK_1}{dx_2} + K_2.$$

Equation (2.36), together with the evaluation of K, forms the basis of the theory for a nonuniform system with the second term in the integral generally being referred to as the gradient energy.[†] The evaluation of $f(x_2, 0, 0, \ldots)$ and of K requires the assumption of a particular model for the solution or a knowledge of the interactions between pairs of atoms.

To evaluate the specific surface free energy f^s we recall the definition

$$Af^s = F^{\text{total}} - (F^\alpha + F^\beta) \tag{2.37}$$

in the terminology of section 1.2. The sum of the fictitious free energies F^α and F^β may be written

$$F^\alpha + F^\beta = A \int N_v \{ x_2' \mu_2 + (1 - x_2') \mu_1 \} \, dz, \tag{2.38}$$

where μ_1 and μ_2 are the chemical potentials of the two components and x_2' is the value of the mole fraction of 2 appropriate to the bulk of each phase; in eqn. (2.38) $x_2' = x_2^\alpha$ for $z \leqslant 0$ and $x_2' = x_2^\beta$ for $z > 0$. Comparing eqns. (2.36) and (2.37) leads to

$$f^s = \int N_v \left\{ \Delta f(x_2, 0, 0, \ldots) + K \left(\frac{\partial x_2}{\partial z} \right)^2 + \ldots \right\} dz \tag{2.39}$$

with

$$\Delta f = f(x_2, 0, 0, \ldots) - \{ x_2' \mu_2 + (1 - x_2') \mu_1 \}. \tag{2.40}$$

The equilibrium distribution of the components through the interface at constant temperature and volume is, of course, the one which corresponds to minimizing the total Helmholtz free energy of the system F. For a particular choice of the dividing surface this is equivalent to minimizing f^s as can be seen from the defining equation (2.37). Using the Euler rule for minimizing the integral expression for f^s leads to

$$\Delta f(x_2, 0, 0, \ldots) = K \left(\frac{\partial x_2}{\partial z} \right)^2 \tag{2.41}$$

[†] In the general case the gradient energy coefficient K will be a second rank tensor and the gradient energy will depend on the orientation of the surface.

as the desired equilibrium condition, and hence

$$f^s = 2 \int N_v \{ \Delta f(x_2, 0, 0, \ldots) \} \, dz$$

$$= 2 \int_{x_2^\alpha}^{x_2^\beta} N_v \{ K \, \Delta f(x_2, 0, 0, \ldots) \}^{1/2} \, dx_2. \tag{2.42}$$

Equation (2.41) is essentially the differential equation for the composition profile through the interface and corresponds to the set of eqns. (2.24) of the previous section. Its solution of course requires the assumption of a particular model for the mixture. We shall not pursue this analysis here but note only that it is clear from eqns. (2.42) or (2.39) with the definition (2.40) that the specific surface free energy f_s is dependent on the location of the dividing surface (the plane $z = 0$) within the transition region as was mentioned at the beginning of the chapter. If the composition profile were obtained, the surface excesses

$$\Gamma_i = \int N_v (x_i - x_i') \, dz$$

could then be determined and also the surface tension

$$\gamma = f^s - \Gamma_1 \mu_1 - \Gamma_2 \mu_2$$

In this chapter we have discussed some thermodynamic aspects of interfaces in multicomponent systems and have attempted to indicate the principles involved in describing analytically the composition variations across interfaces between phases in binary systems. We have done this by outlining two possible approaches—one a discrete layer model which will be most appropriate when rapid composition changes are involved, and the other a continuum model which will be a good approximation when the composition varies very slowly with position through the interface. It would be rather satisfying at this point to be able to compare theory and experiment on the structure of interfaces in binary and other alloys. However, the theoretical side of the problem is still being developed[†] and few quantitative predictions have

---

[†] The validity of the Cahn–Hilliard approach is discussed in considerable detail in a recent series of papers by R. Kikuchi, Boundary free energy in the lattice model, *J. Chem. Phys.* **20**, 30 (1972).

yet been made. There are also few experimental data although this situation is likely to change as rapid progress is presently being made on the development of new techniques for surface compositional analysis. A few systems have recently been studied in which marked segregation of one component occurs. For example, certain surfaces of dilute Ni–C alloys show sharp variations of the surface C concentration with temperature. As experimental data become available, advances in the theory may also be expected.

BIBLIOGRAPHY

ADAMSON, A. W., *Physical Chemistry of Surfaces*, 2nd edn., Interscience, New York, 1967.

CAHN, J. W., and HILLIARD, J. E., Free energy of a nonuniform system: I, Interfacial free energy, *J. Chem. Phys.* **28,** 258 (1958).

GUGGENHEIM, E. A., *Mixtures*, Oxford University Press, 1952.

HART, E., Thermodynamic functions for nonuniform systems, *J. Chem. Phys.* **39,** 3075 (1963).

HONDROS, E. D., and MCLEAN, D., Surface energies of solid metal alloys, in *Surface Phenomena of Metals*, Society of Chemical Industry, Monograph No. 28, London, 1968.

MEIJERING, J. L., Concentrations at interfaces in binary alloys, *Acta Met.* **14,** 251 (1966).

MURAKAMI, T., ONO, S., TAMURA, M., KURATA, M., On the theory of the surface tension of a regular solution, *J. Phys. Soc. Japan* **6,** 309 (1951).

SEMENCHENKO, V. K., *Surface Phenomena in Metals and Alloys*, Pergamon, Oxford, 1962.

STILLINGER, F. H., and BUFF, F. P., Equilibrium statistical mechanics of inhomogeneous fluids, *J. Chem. Phys.* **37,** 1 (1962).

CHAPTER 3

EXPERIMENTAL MEASUREMENTS OF SURFACE TENSION IN SOLIDS

3.1. INTRODUCTION

The measurement of the surface tensions of solids presents considerably more difficulty than in the case of liquids for which a variety of accurate methods, notably the sessile drop technique, have been developed. Most of the additional problems in the case of solids are associated with the low atomic mobilities characteristic of the solid state, and this results in extremely long times for the attainment of equilibrium surface shapes. The amount of reliable data on solid surface tensions is still relatively meagre, and it is often necessary to obtain a rough estimate for the solid from the value for the liquid at the melting point. This latter procedure has some theoretical justification in that the local coordination and type of bonding in liquids, particularly metals, is normally not greatly different from that of the solid state.

The most direct method for the determination of surface tensions in solids is one which involves the measurement of the forces necessary for the propagation of cleavage cracks. This technique is, of course, limited to brittle solids. For metals the most generally useful method appears to be the so-called "zero creep" method in which one measures the forces necessary to counteract the tendency of a fine wire or sheet to decrease its surface area due to the surface tension. Both of these methods are in principle absolute since they involve only the measurement of external forces. Values of surface tension can also be obtained from observations on the rate of mass transport which is induced by capillarity. The method is somewhat more difficult experimentally and not of great accuracy since the surface tension values so determined depend on the accuracy to which diffusion coefficients are known. Finally, a method which so far has met only with limited success is that based on comparison of the heat of solution of finely divided material of large surface area with that determined for normal bulk material. We shall consider the main features of these experimental measurements as well as techniques for the detection of surface stress and the variation of surface tension with orientation.

3.2. SURFACE TENSION FROM CLEAVAGE EXPERIMENTS

Since surface tension is by definition the reversible work of creating unit area of new surface, the measurement of the work involved in cleaving a crystal to expose two new surfaces approaches most closely the situation envisaged by theorists in computing values of γ. The method was apparently first used by Obreimoff and by Orowan in connection with experiments on mica and has more recently been developed by Gilman and others. The type of experimental set up used is indicated schematically in Fig. 3.1. With a force F applied as

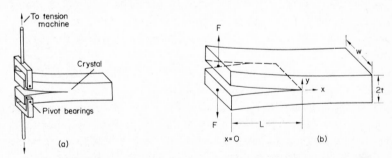

FIG. 3.1. Diagram illustrating the measurement of surface tension by the cleavage technique as used by Gilman. The force F applied at the end of the crystal is determined for which the decrease in the elastic stored energy on propagating the crack becomes equal to the corresponding increase in free energy due to the creation of new surface. (From J. J. Gilman, *J. Appl. Phys.* **31**, 2208 (1960).)

shown in the figure, the system will be stable so long as the decrease in the elastic energy stored in the crystal on propagating the crack is less than the corresponding increase in free energy accompanying the creation of the two new surfaces. The load F is slowly increased until the point of instability is reached. Approximating the stored energy as that in two cantilever beams, the surface tension is related to the critical load by

$$\gamma = 6 \frac{F^2 L^2}{E w^2 t^3},\tag{3.1}$$

where E is Young's modulus of the crystal, L the initial crack length and w and t the width and thickness of the crystal respectively.

The main difficulty in the interpretation of the results of this experiment arises from the irreversible nature of the cleavage process. In order for the process to approach reversibility it would have to occur at extremely small velocity and under ambient conditions such that no gas adsorption occurs. In addition, dissipative processes will of course make the process irreversible; these include plastic flow, the propagation of acoustic waves from the moving crack tip, the formation of point defects and dislocations, and heat flow to the surroundings. To decrease the effect of dislocation motion some experiments have been done at low temperatures. An indication of the effect of temperature can be seen in the results on Fe–3% silicon single crystals; for cleavage on the (100) plane the effective values of the surface tension obtained from eqn. (3.1) were 1360 ergs/cm² at 20°K, 24,500 ergs/cm² at 77°K, and 240,000 ergs/cm² at 195°K. It is thus obvious that the amount of energy dissipated as heat and stored as imperfections in the crystal can greatly exceed the energy stored as surface free energy. For more brittle materials more consistent results have been obtained and these are summarized in Table 3.1. The method is of great interest since in principle it provides an absolute value of γ and is most applicable in the low temperature region where other methods are inappropriate. Refinements immediately suggested by the work to date are the use of liquid helium temperatures and ultra high vacuum conditions. A more detailed study of the mechanics of the experimental

TABLE 3.1. COMPARISON OF ESTIMATED SURFACE TENSIONS
WITH VALUES FROM CLEAVAGE EXPERIMENTS

Crystal	Surface tension (ergs/cm²)	
	Experimental value	Estimated from equation (5.15)
LiF (100)	340	370
MgO (100)	1200	1300
CaF$_2$ (111)	450	540
BaF$_2$ (111)	280	350
CaCO$_3$ (1010)	230	380
Si (111)	1240	890
Zn (0001)	105	185
Fe–3 % Si (100)	~1360	1400

The value of d_0 in eqn. (5.15) was taken to be the distance between the crystallographic cleavage planes; a was taken as the mean atomic radius of atoms in the cleavage plane. (From data given by J. J. Gilman, *J. Appl. Phys.* **31**, 2208 (1960).)

tuation shows that eqn. (3.1) should be modified to

$$\frac{1}{\gamma} = \frac{Ew^2t^3}{6F^2L^2} - c\frac{t^2}{L^2}, \qquad (3.2)$$

where c is a constant characteristic of the crystal. According to the more detailed analysis, the limit $(t/L) = 0$ corresponds to the situation when there is no plastic zone ahead of the propagating crack tip. Thus measurements of F for different values of the geometrical parameters should be capable of eliminating the contribution from plastic deformation.

The question has been raised as to whether the value obtained for γ from such cleavage experiments applies to the final equilibrium state of the surface or to some intermediate configuration through which the system passes. Thus we may picture the process occurring in two steps: (a) the creation of two new surfaces in which the configurations of the atoms are the same as in corresponding planes in the bulk, and (b) small relaxations of the surface atoms to new positions with the release of energy. If f_a and f_b are the increases in free energy in these two steps per unit area of surface, then

$$\gamma = f_a + f_b, \tag{3.3}$$

where the contribution f_b will, of course, be negative. Since experimentally the process is observed to be irreversible, it is possible that the quantity measured is f_a and this may be significantly greater than γ. There is, in fact, some indication that the experimental value may be in better agreement with surface energy calculations which ignore lattice relaxation. It should also be mentioned that the surface tension for a surface created by cleavage may exceed that for the equilibrium surface configuration since it is possible that the minimum free energy condition may only be achieved by rearrangements which involve the migration of atoms within the crystal or over the surface; these processes do not occur at any significant rate at the temperatures of the cleavage experiments.

3.3. METHOD OF ZERO CREEP

This technique appears to be most suitable for those materials in which the surface tension is not a strong function of orientation and which can be fabricated in the form of thin sheets or wires a few mils in diameter. Most of the cubic metals satisfy these criteria. The measurements are in this case carried out at high temperature where atoms are fairly mobile, and it is generally agreed that the surface tension and surface stress will be equal under these experimental conditions.

Figure 3.2 shows the type of specimen used in the experiment; it is a polycrystalline wire with a bamboo-like structure due to the formation, during annealing, of grooves at the intersections of the horizontal grain boundaries with the cylindrical surface of the wire [described by eqn. (1.18)]. The experiment consists in finding the applied stress necessary to prevent either shrinkage or elongation of the wire

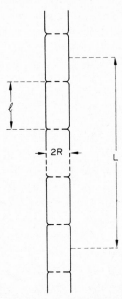

FIG. 3.2. Form of the specimen used in the zero creep experiment for determining surface tensions of solids. It consists of a fine wire with a diameter of a few mils with a so-called bamboo structure due to the formation of grooves at the positions of the grain boundaries during annealing. The quantities used in eqns. (3.4) and (3.5) are indicated.

as determined by observing a section of the wire while at high temperatures with various loads. If the mass of the section may be neglected in comparison with the external load, the equilibrium condition will be obtained when the gravitational work done on the external load for any small displacement is equal to the increase in the quantity γA. In fact a small correction usually has to be made for the change in area of the grain boundaries within the section, and the external force for equilibrium is given by

$$F = \frac{d}{dl}(2\pi RL\gamma + \pi R^2 \gamma_g N_g), \qquad (3.4)$$

where R is the wire radius, γ_g the grain boundary tension, and N_g the number of grain boundaries within the section of length L. Taking the volume of the section to be fixed as well as N_g, we obtain the load required for equilibrium

$$F = \pi R \gamma \left\{ 1 - \frac{\gamma_g}{\gamma} \left(\frac{R}{l} \right) \right\}$$

or an applied stress

$$\sigma = \frac{\gamma}{R} \left\{ 1 - \left(\frac{\gamma_g}{\gamma} \right) \left(\frac{R}{l} \right) \right\}. \tag{3.5}$$

The quantity γ_g / γ can be estimated by measuring the dihedral angles at the grain boundary–surface intersections and using eqn. (1.18) with the orientation derivatives neglected. Figure 3.3 shows meas-

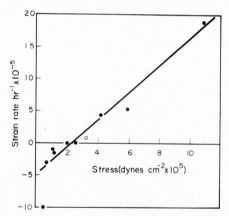

Fig. 3.3. Strain rate versus applied stress for polycrystalline nickel wires in argon at 1420°C. The surface tension is calculated from the applied stress which corresponds to zero creep. (From E. R. Hayward and A. P. Greenough, *J. Inst. Metals* **88**, 217 (1960).)

urements of the strain rate of fine polycrystalline nickel wires with applied loads showing both extension and contraction ranges. It should be emphasized that since the value of γ is calculated from the zero creep condition it is not dependent on the particular model we may choose to describe the creep process. It is in principle an absolute determination of γ. The mechanism of creep under these circumstances is, nevertheless, of some interest, since in order to determine the equilibrium condition with reasonable accuracy, significant creep rates have to be achieved. In this connection it is interesting to note that single crystal wires subjected to similar stresses show no appreciable creep. It thus appears that although the presence of the grain boundaries necessitates a correction factor in eqn. (3.5) they are necessary for the success of the technique. Figure 3.4 depicts a probable mechanism of shape change of the individual grains; this is the Nabarro–Herring mechanism in which there is a net flow of atoms between

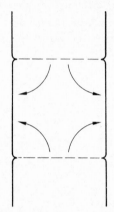

FIG. 3.4. Probable mechanism of mass transport causing shape change of individual grains due to an applied load. The arrows indicate the direction of vacancy flow during extension (or atom flow during contraction), i.e. from the grain boundaries to the free surface. A high density of grain boundaries is necessary to achieve significant creep rates. This is known as the Nabarro–Herring mechanism of creep. (After C. Herring, *J. Appl. Phys.* **21,** 437 (1950).)

the free surface and the grain boundaries either by the process of volume diffusion or by surface diffusion. It is clear that the rate of shape change (and hence creep rate) is dependent on the dimensions of the individual grains, and, in fact, volume diffusivities can be derived from observation on creep rates.

We note that since the zero creep measurements are made at high temperatures, the resulting value of γ will refer to equilibrium situations. This means that the technique is appropriate for studying the effect of equilibrium solute segregation or gas adsorption on surface tension; on the other hand, small impurity concentrations may be rather important with this technique whereas they should be relatively unimportant in the cleavage type of measurement. Figures 2.3 and 2.4 represent some results of this type; Fig. 2.3 refers to the effect of O_2 adsorption on the surface tension of silver from measurements on polycrystalline wires while Fig. 2.4 indicates the influence of antimony segregation to copper surfaces; the latter results were obtained from experiments using specimens in the form of thin-walled hollow cylinders instead of wires.

Another interesting method which is similar in principle to that of zero creep is one involving the use of field emission microscopy specimens (see section 6.2.4). A field emission microscope specimen consists of a sharp tip (Fig. 3.5) of radius about 1×10^{-5} cm which has been produced by electropolishing and by heating and cleaning in vacuum. Such a specimen is morphologically unstable and will tend to blunt at sufficiently high temperatures, a process which can in fact be used in the study of surface self-diffusion. If, however, a sufficiently high voltage is applied to the tip (~ 30 kV) relative to its surroundings it may be possible to counterbalance the chemical potential gradients, which result from differences in curvature, by electrostatic energy differences. It can readily be shown that for a perfect conductor in an electrostatic field the relationship corresponding to eqn. (1.27) must include a term representing the electrostatic stored energy and becomes

$$\Delta\mu = \Omega_0 \left\{ \gamma\left(\frac{1}{R_1} + \frac{1}{R_2}\right) + \left(\frac{\partial^2\gamma}{\partial n_x^2}\frac{1}{R_1} + \frac{\partial^2\gamma}{\partial n_y^2}\frac{1}{R_2}\right) - \frac{F^2}{8\pi} \right\}, \quad (3.6)$$

where F is the local value of the applied electrostatic field at the portion of the surface of radii of curvature R_1 and R_2. The solution for F in terms of the applied voltage requires a detailed knowledge of the specimen geometry. The experimental method used for the detection of the blunting process consists in the measurement of the rate of removal (by diffusion) of surface steps, Fig. 3.5(b, c), as seen in the

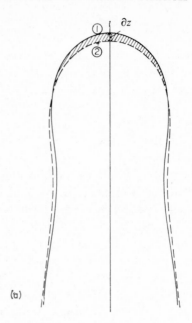

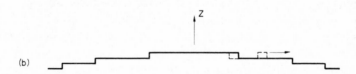

FIG. 3.5. (a) Approximate shape of a field emission microscope specimen. The radius of the tip is generally 1000–10,000 Å. On annealing, blunting occurs by diffusion away from the tip to the shank of the specimen. The dotted outline ② indicates the shape after a short annealing period. (b) This indicates the type of step array near the apex. Blunting may occur by the detachment of atoms from the steps and subsequent diffusion over the surface which leads to a shrinkage of the step radii. The steps appear as bright rings on the emission pattern due to the enhanced electron emission near the step. The shrinkage rate can be obtained from the measured rate of ring collapse.

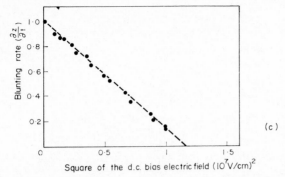

FIG. 3.5. (c) Plot of the blunting rate measured by the rate of recession, $\partial z/\partial t$, of the topmost plane (normalized to the zero field value) against F^2. (Figures 3.5(a) and (c) from J. P. Barbour *et al.*, *Phys. Rev.* **117**, 1452 (1960).)

electron emission image. It is interesting to note that the reverse process to blunting does not occur even at very high fields, presumably because of the necessity of nucleating new surface layers; instead a process termed "build-up" is observed in which the tip generally assumes a polyhedral shape.

The value of γ is obtained from eqn. (3.6) by ignoring the orientation dependence, i.e.

$$2\gamma = \frac{R}{8\pi}F_0^2, \tag{3.7}$$

where F_0 is the extrapolated field at the emitter tip for which the recession rate becomes zero. The technique has apparently been applied only to the refractory metals which can be held at a sufficiently high temperature during the measurements to prevent gas adsorption.

3.4. MASS TRANSPORT MEASUREMENTS

As indicated in section 1.6, the variation of curvature at the surface of a solid will lead to gradients in chemical potential and this will in turn produce atomic fluxes at sufficiently high temperatures. In prin-

TABLE 3.2. SOME EXPERIMENTAL VALUES OF SURFACE
TENSION FOR METALS

Material	Temperature (°K)	Surface tension γ (ergs/cm^2)[a]
Ag	1180	1140
Au	1300	1410
Be	973	810
Cu	1320	1670
δ-Fe	1750	1950
γ-Fe	1700	2150
Mo	2623	1960
Nb	2520	2100
Ni	1523	1850
Ni (100)	1490	1821
Ni (110)	1490	1900
Pt	1520	2340
W	2000	2900
Zn	650	830

[a] 1 erg/cm^2 = 10^{-3} joule/m^2.

Unless otherwise indicated the values listed represent
an average over a range of surface orientations. References to the individual measurements may be found in the
article by J. M. Blakely and P. S. Maiya in *Surfaces and
Interfaces I* (ed. J. J. Burke, N. L. Reed, and V. Weiss),
Syracuse University Press, 1967, p. 325, or in the article
by G. Ehrlich in *Surface Phenomena of Metals*, Society
of Chemical Industry, Monograph No. 28, 1968, p. 13.

ciple the rate of the resulting mass transport can be used to determine
the surface tension when all other parameters are known. We shall
discuss the phenomenon of mass transport in some detail in section
7.3. For the sake of illustration consider for the moment the rate of
approach to planarity of a surface which has a sinusoidal corrugation
(see Figs. 7.5 and 7.6 (pp. 215 and 225)). When volume and surface
diffusion are responsible for the decay the amplitude decays exponentially with time, at fixed temperature, with a decay constant α given by

$$\alpha = B' \frac{D_s \gamma}{\lambda^4} + C' \frac{D_B \gamma}{\lambda^3}, \tag{3.8}$$

where λ is the wavelength of the corrugation, D_B and D_s are volume and surface diffusion coefficients appropriate to mass transport processes respectively, and B' and C' are constants for a particular crystal (see section 7.3). It is clear that γ may be obtained using eqn. (3.8) from an experimental determination of α if all other parameters are

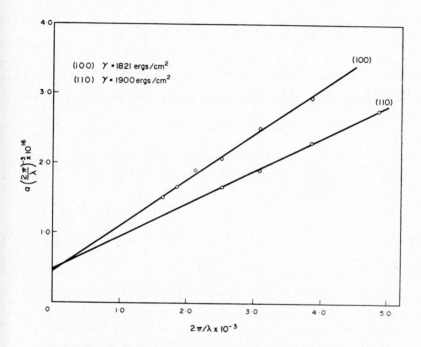

FIG. 3.6. Plot of $\alpha(2\pi/\lambda)^{-3}$ versus $2\pi/\lambda$ for the determination of surface tension from the rate of decay of sinusoidal surfaces of wavelength λ. α is the amplitude decay constant. The results given here are for surfaces of nickel near (100) and (110) at $\sim$ 1220°C. (From J. M. Blakely and P. S. Maiya, in *Surfaces and Interfaces I*, ed. J. J. Burke, N. L. Reed, and V. Weiss, Syracuse University Press, 1967, p. 325.)

known. Measurements have in fact been made of α as a function of λ for nickel and platinum (included in Table 3.2) and γ determined from the intercept of a plot of $(\alpha\lambda^3)$ versus $1/\lambda$ (Fig. 3.6). The method is of interest from the point of view of studying single crystal surfaces and provides values of γ for a small range of orientation. It is, however, not an accurate method for the absolute determination of γ since it involves the use of a volume diffusion coefficient whose absolute magnitude is generally not known with high precision.

3.5. CALORIMETRIC MEASUREMENTS

For a one-component system we have asserted that we may identify the surface tension γ with the specific surface Helmholtz free energy f^s, and from section 1.2

$$f^s = e^s - Ts^s,$$

where e^s and s^s are the specific surface internal energy and entropy respectively. Using the concept of the dividing surface $e^s = h^s$, the specific surface enthalpy, so that

$$\gamma = f^s = h^s - T \int_0^T \frac{C_P^s \, dT}{T}, \qquad (3.9)$$

where C_P^s is the excess heat capacity per unit area of surface. It is thus apparent that surface tension could be computed from calorimetric measurements of surface enthalpy and excess heat capacity, quantities which are themselves of considerable fundamental interest in connection with bonding at the surface and surface atom vibrations. Attempts to obtain these quantities have involved the comparison of the heat of solution and heat capacity of fine powders with the corresponding quantities for bulk materials. In order for the surface to make a significant contribution to the experimental observations it has been estimated that the mean particle size must be less than about 10^{-4} cm, a requirement which leads to considerable uncertainty in the results. This is due to the errors in surface area estimation, the range of different morphologies involved, and—perhaps most importantly—the fact that it is extremely difficult to devise an arrangement to produce fine powders with uncontaminated surfaces. This last point is probably not quite so important with the oxides and alkali halides with which the measurements have so far been made as it would be with metallic powders for which heats of adsorption of oxygen and other gases are very high. The interpretation of the results that have been obtained so far is very uncertain, and we shall not pursue the topic here; the data are discussed in the article by Benson and Yun cited in the bibliography. Since techniques are presently being developed for studying surface vibrations it is likely that there will be renewed interest in the experimental determination of excess surface heat capacity.

3.6. MEASUREMENT OF SURFACE STRESS IN SOLIDS

Most of the measurements of surface stress in solids have been based on the interpretation of density changes for small particles through the equation of Young and Laplace, eqns. (1.19). This strictly applies only to spherical particles in cases where the surface stress is isotropic but it has, nevertheless, been used to give approximate interpretations of results on polyhedral crystalline particles. The experimental results are still rather sparse and are somewhat inconsistent probably for the reasons mentioned in the previous section. Among the more recent observations is some work on small particles of gold produced by aggregation of thin films and examined by electron diffraction methods in the electron microscope. At 50°C Mays *et al.* find lattice parameter changes of the order of 0·01 Å in particles of diameter ~ 50 Å corresponding to a surface tensile stress (obtained from $\Delta a/a = \frac{1}{3}\beta(2g/R)$, eqn. (1.22)) of 1175 ± 200 dyne/cm. An extrapolation of the surface tension measured by the zero creep method at high temperatures gave a value of γ of ~ 1800 ergs/cm^2 at 50°C, giving support to our earlier assertion that at relatively low temperatures in solids the surface tension and surface stress will in general be unequal.

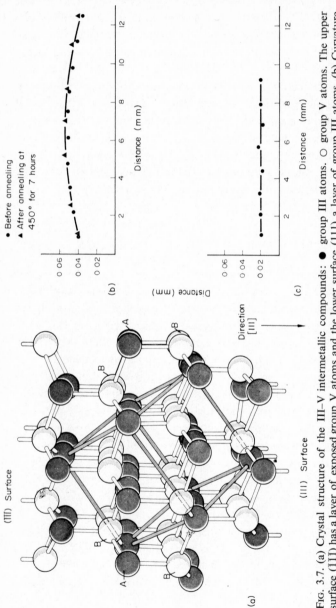

FIG. 3.7. (a) Crystal structure of the III–V intermetallic compounds: ● group III atoms. ○ group V atoms. The upper surface (11̄1̄) has a layer of exposed group V atoms and the lower surface (111) a layer of group III atoms. (b) Curvature of an InSb {111} wafer 7.8 μ thick. The fact that no significant difference is produced by annealing indicates that the effect is due to the intrinsic properties of the surfaces. (c) The results of a control experiment on a {111} Ge wafer of the same thickness. (From H. C. Gatos in *Surface and Interfaces I*, ed. J. J. Burke, N. L. Reed, and V. Weiss, Syracuse University Press, 1967.)

Another interesting demonstration of the existence of surface stress has been made for InSb and GaAs crystals. These crystals have the cubic zinc sulfide structure which may be viewed as a layered structure consisting of alternate layers of group III and group V atoms and are non-centrosymmetric. They may be cleaved on the close-packed planes and the two surfaces (111) and ($\bar{1}\bar{1}\bar{1}$), say, which are exposed differ in that one has, in the case of InSb, an outer layer of indium while the other has an outer layer of antimony [Fig. 3.7(a)]. It has in fact been found that the two surfaces have different chemical activities and electrical properties. On forming a thin sheet bounded by these close-packed planes the difference in the surface stress of the two surfaces causes a bending as indicated in Fig. 3.7(b), and this surface stress difference may be obtained from measurements of the curvature. For example, for InSb the observations have been analyzed to give a surface stress difference of about 10^3 dyne/cm between the two surfaces. An interesting feature of these observations was that the amount of bending, and hence the surface stress difference, could be altered by exposure to different gases.

3.7. DETERMINATION OF THE VARIATION OF γ WITH ORIENTATION

Three main methods have been used to obtain experimental information on the Wulff plot of crystals. These are: (1) the measurement of dihedral angles at twin or grain boundary–surface intersections, (2) the study of the equilibrium shapes of small particles, voids, or bubbles, and (3) observations on faceting of planar surfaces. Of these only method (1) is capable of yielding the entire γ-plot while the others are usually restricted to giving ratios of the surface tensions for different orientations.

3.7.1. *Twin and Grain Boundary–Surface Intersections*

This method was first used by Mykura who observed and interpreted the types of morphologies shown in Fig. 1.12 (p. 23) at the intersections of pairs of twin boundaries with free surfaces in face-centered cubic metals. As indicated in section 1.5, these morphologies can only be explained by considering the variation of surface tension with orientation. The equations corresponding to eqn. (1.16) for the pair of twin boundary–surface intersections are, with the notation of Fig. 1.12(c),

$$\gamma_T = \gamma_Q \cos A + \gamma_R \cos B - \frac{\partial \gamma_Q}{\partial A} \sin A - \frac{\partial \gamma_R}{\partial B} \sin B,$$

$$\gamma_T = \gamma_{Q'} \cos A' + \gamma_{R'} \cos B' + \frac{\partial \gamma_{Q'}}{\partial A'} \sin A' + \frac{\partial \gamma_{R'}}{\partial B'} \sin B',$$

(3.10)

where γ_T is the surface tension associated with the twin boundary, usually referred to as the twin boundary tension. Equations (3.10) are derived from eqn. (1.16) by considering the components parallel to the twin boundary, the assumption being that the twin cannot rotate out of its {111} type of orientation because of the rapid increase in boundary tension with orientation in this case. The measurements made are of the angles in eqn. (3.10) and also the crystallographic orientations of the surfaces involved. It appears that the most reliable way of constructing the γ-plot from such measurements is to express the γ-plot as a double Fourier series with a finite number of unknown coefficients a_{mn}, b_{mn}, c_{mn}, and d_{mn}, i.e.

$$\frac{\gamma(\theta, \phi)}{\gamma_T} = \sum_{m, n} (a_{mn} \cos m\theta \cos n\phi + b_{mn} \cos m\theta \sin n\phi$$

$$+ c_{mn} \sin m\theta \cos n\phi + d_{mn} \sin m\theta \sin n\phi), \quad (3.11)$$

where θ, ϕ are the angles between the surface normal and some reference crystallographic direction, and to substitute eqn. (3.11) in each of eqns. (3.10) to obtain a linear equation in the unknown coefficients. With a sufficient number of observations these equations can be solved for a fairly large number of coefficients, and hence a realistic γ-plot can be constructed. Figure 3.8 shows an example of a γ-plot for gold in a hydrogen atmosphere at 1030°C determined in this way. So far such measurements have been made for a number of metals including

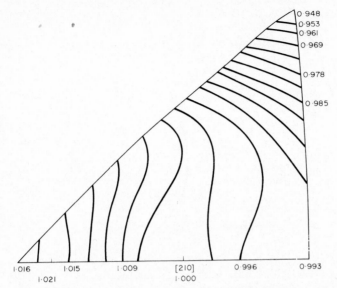

FIG. 3.8. Representation in the stereographic triangle of the normalized γ-plot of Au in a hydrogen atmosphere at 1030°C determined by the twin boundary–surface intersection method. The contours represent orientations of equal surface tension. The variation over all orientations appears to be about 5%. (Courtesy W. L. Winterbottom and N. A. Gjostein.)

Ni, Cu, Au, γ-Fe, Co, and Pt although most of the data have been analyzed by a more approximate method than that outlined above. It appears from these measurements that for the atmospheres used there are indeed minima at the more closely packed surfaces with an overall variation in surface tension of about 10% or so.

As indicated, the method relies on the frequent occurrence of pairs of annealing twins in polycrystalline specimens, a phenomenon which is most common in face-centered cubic metals which have low twin boundary tensions. For materials in which twinning is rare, such as the body-centered cubic metals and most compounds, a more appropriate method was described fairly recently by McLean and Gale (see bibliography) involving the measurement in the electron microscope of the variation in dihedral angle at grain boundary grooves around the circumference of fine polycrystalline wires with a "bam-

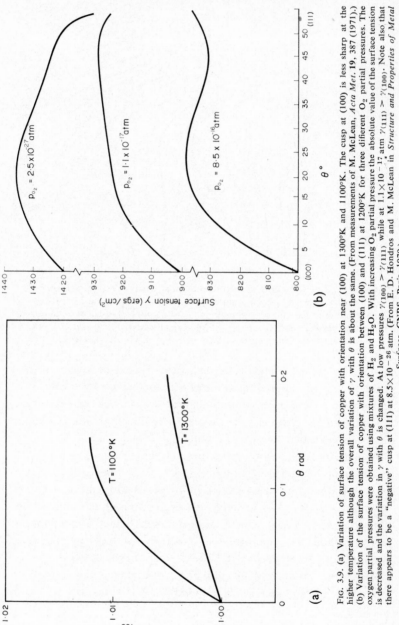

Fig. 3.9. (a) Variation of surface tension of copper with orientation near (100) at 1300°K and 1100°K. The cusp at (100) is less sharp at the higher temperature although the overall variation of γ with θ is about the same. (From measurements of M. McLean, *Acta Met*, **19**, 387 (1971).) (b) Variation of the surface tension of copper with orientation between (100) and (111) at 1200°K for three different O₂ partial pressures. The oxygen partial pressures were obtained using mixtures of H₂ and H₂O. With increasing O₂ partial pressure the absolute value of the surface tension is decreased and the variation in γ with θ is changed. At low pressures $\gamma_{(100)} > \gamma_{(111)}$ while at 1.1×10^{-17} atm $\gamma_{(111)} > \gamma_{(100)}$. Note also that there appears to be a "negative" cusp at (111) at 8.5×10^{-26} atm. (From E. D. Hondros and M. McLean in *Structure and Properties of Metal Surfaces*, CNRS, Paris, 1970.)

boo" type of structure (Fig. 3.2) and using essentially the same type of analysis as in the twin boundary method. This technique is apparently considerably less tedious than the twin boundary method and has been used for the study of the variation of the γ-plot with temperature and with pressure of an adsorbing gas. Figure 3.9(a) shows some results on the shape of the γ-plot over a section near the (100) orientation for copper at two different temperatures. The results indicate that with increasing temperature the sharpness of the minimum tends to decrease, a trend to be expected from considerations of the thermal creation of surface defects as discussed in Chapter 4. As the temperature is raised, entropy considerations suggest that the density of atomic defects on singular and other surfaces will increase, and it is possible that it becomes sufficiently high that the difference in the structure of surfaces of different orientations essentially disappears. Figure 3.9(b) shows the effect of gas adsorption on a section of the γ-plot of copper; the different partial pressures of oxygen were produced by using H_2O and H_2 mixtures. As well as decreasing the average value of γ the adsorption leads to changes in the shape of the γ-plot. At sufficiently low pressures of oxygen $\gamma_{(111)} < \gamma_{(100)}$ while the relative magnitudes are reversed at $\sim 10^{-17}$ atm of O_2. There appears also to be an outward pointing cusp in the γ-plot at (111) for sufficiently high pressures.

It is possible to construct a theory for the shape of γ-plot cusps based on the use of models of surfaces with arrays of atomic steps which are assumed to contribute to the energy and entropy of the surface and to interact with each other. We shall consider this microscopic model in Chapter 4. As more extensive data on γ-plots become available it should be possible to evaluate various parameters of the theoretical description, in particular the energies and entropies associated with steps and also the changes in these quantities which accompany adsorption.

3.7.2. *Equilibrium Shapes and Faceting*

The fact that most small crystalline particles take on polyhedral forms when annealed under equilibrium conditions is perhaps the most direct demonstration of the existence of γ-plot cusps. Figure 3.10(a) is an electron micrograph of replicas of silver particles annealed on a substrate in vacuum and shows well-developed facet planes which can be identified as low index planes from the symmetry prop-

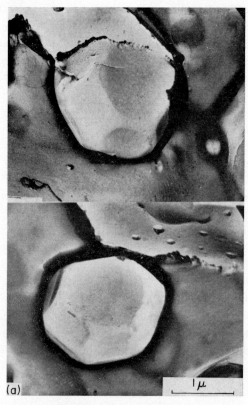

(a)

1μ

FIG. 3.10(a). Electron micrograph replicas of silver particles annealed on a substrate in vacuum at 700°C for 100 hr. The polyhedral shape is believed to be directly related to the γ-plot. (Courtesy of B. E. Sundquist.)

erties. However, problems arise in deriving quantitative information on the surface tension variation from such observations due to the presence of external forces connected with supporting the particles, the possibility of interaction with the substrate, and the difficulty of maintaining clean surfaces for small particles. It is possible, however, by measuring the relative areas of different planes, to derive approximate surface tension ratios for different orientations. For example, from observations on silver particles which have been annealed at 700°C on a BeO substrate it has been found that $\gamma_{(111)}/\gamma_{(100)} \simeq 0.95$ and $\gamma_{(100)}/\gamma_{(110)} \simeq 0.91$.

Also related to the γ-plot are the faceted equilibrium shapes often assumed by internal voids and bubbles (Fig. 3.10(b)) which may be produced in crystals, e.g. by irradiation with neutrons, α-particles, or fission products, and in fact with such internal surfaces the equilibrium shape is more readily related to the free surface γ-plot since there is no contact with another surface.

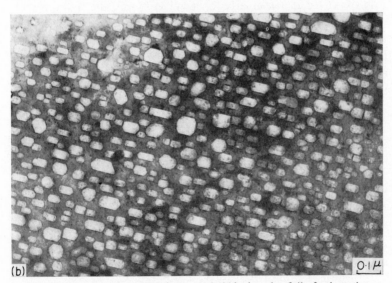

(b)

FIG. 3.10(b). Micrograph showing argon bubbles in a zinc foil of orientation near [11$\bar{2}$0]. The foil was irradiated with 60 keV argon ions (60 μA-min) and annealed for 1 hr at 230°C. (Courtesy of G. A. Chadwick.)

The conditions for stability of a planar surface with respect to faceting have been discussed in section 1.4. We have also seen that when faceting does in fact occur, the angles between the new surfaces can be used to give relative values of their surface tensions and also some information on the slope of the γ-plot near the cusp. A great many observations of faceting have now been reported in the literature from a variety of experimental methods including optical microscopy, low and high energy electron diffraction, and field emission microscopy. The question of whether the faceted configuration is the equilibrium one or whether it results from anisotropies in evaporation rate has not really been determined in many cases. However, with the assumption that the morphologies are in fact equilibrium ones, it is found, for example, that in silver (the system studied most extensively) in contact with air at about 900°C the ratio of the surface tension of (100) to that of general high index surfaces is about 0.90 with a corresponding value for (111) of about 0.84.

It appears that particularly for metallic crystals the methods mentioned in section 3.7.1 are likely to yield the most extensive information on γ-plots. Similar studies with other types of crystals, particularly oxides and halides, would be of considerable interest. Microscopic considerations suggest that the anisotropies in these crystals would be much greater than is found for metals.

BIBLIOGRAPHY

BARBOUR, J. P., CHARBONNIER, F. M., DOLAN, W. W., DYKE, W. P., MARTIN, E. E., and TROLAN, J. K., Determination of the surface tension and surface migration constants for tungsten, *Phys. Rev.* **117**, 1452 (1960).

BASTERFIELD, J., MILLER, W. A., and WEATHERLY, G. C., Anisotropy of interfacial free energy in solid–fluid and solid–solid systems, *Can. Metall. Quart.* **8**, 131 (1969).

BENSON, G. C., and YUN, K. S., Surface energy and surface tension of crystalline solids, in *The Solid–Gas Interface*, Vol. 1 (ed. E. A. Flood), Dekker, New York, 1967, chap. 8.

BLAKELY, J. M., and MAIYA, P. S., Surface energies from transport measurements, in *Surfaces and Interfaces I* (ed. J. J. Burke, N. L. Reed, and V. Weiss), Syracuse University Press, 1967.

GILMAN, J. J., Direct measurements of the surface energies of crystals, *J. Appl. Phys.* **31**, 2208 (1960).

HONDROS, E. D., Surface energy measurements, in *Techniques in Metals Research*, Vol. IV, Part 2 (ed. R. A. Rapp), Wiley, 1970.

INMAN, M. C., and TIPLER, H. R., Interfacial energy and composition in metals and alloys, *Metall. Rev.* **8**, 105 (1963).

JACCODINE, R. J., Surface energy of germanium and silicon, *J. Electrochem. Soc.* **110**, 524 (1963).

MCLEAN, M., Determination of the surface energy of copper as a function of crystallographic orientation and temperature, *Acta Met.* **19**, 387 (1971).

MAYS, C. W., VERMAAK, J. S., and KUHLMANN-WILSDORF, D., On surface stress and surface tension: II, Determination of the surface stress of gold, *Surface Sci.* **12**, 134 (1968).

MOORE, A. J. W., Thermal faceting, in *Metal Surfaces*, American Society for Metals, Cleveland, 1963.

OBREIMOFF, J. W., The splitting strength of mica, *Proc. Roy. Soc.* A, **127**, 290 (1930).

UDIN, H., Measurement of solid:gas and solid:liquid interfacial energies, in *Metal Interfaces*, American Society for Metals, Cleveland, 1952, p. 114.

WESTWOOD, A. R. C., and HITCH, T. T., Surface energy of {100} potassium chloride, *J. Appl. Phys.* **34**, 3085 (1963).

WINTERBOTTOM, W. L., Crystallographic anisotropy in the surface energy of solids, in *Surfaces and Interfaces I* (ed. J. J. Burke, N. L. Reed, and V. Weiss), Syracuse University Press, 1967.

CHAPTER 4

ATOMIC STRUCTURE OF CRYSTAL SURFACES

4.1. INTRODUCTION

Our knowledge of the periodic arrangements of atoms in crystals is derived mainly through the technique of X-ray diffraction. It is reasonable to suppose that this regularity is preserved at least to within a small distance of the crystal surface, but it is likely that some distortion of the unit cell may occur in the surface region and there may even be quite drastic structural changes. Because of the high penetrating power of X-rays they are of little value in probing the structure of the outermost planes, and at present most of our knowledge of the atomic structure of surfaces has been derived from experiments using electrons of low energy ($\lesssim 500$ eV) which are strongly attenuated in the crystal and hence give diffraction patterns characteristic of the surface region. Similar observations can be made with higher energy electrons (~ 30 keV) at grazing incidence. These techniques have largely indicated that (with a few notable exceptions such as the (111) surface of silicon)[†] the arrangement in the outermost planes of *clean* surfaces is not radically different from that in parallel atomic planes in the bulk of the crystal. However, the theory of the diffraction of low energy electrons is not yet sufficiently well worked out that fine structural detail can be derived from the observations. Electron diffraction as applied to surfaces has had its greatest success

[†] See, for example, the article by F. Jona, Low energy electron diffraction study of epitaxy of Si on Si, in *Surfaces and Interfaces* (ed. J. J. Burke *et al.*), Syracuse University Press, 1967.

in the study of adsorption where a multitude of new surface phases have been discovered and in observing the initial stages of oxidation and epitaxial growth. The technique of field ion microscopy which has "atomic resolution" is also potentially capable of providing information on various aspects of surface structure.

In discussing many surface phenomena it is useful to visualize a particular model of the atomic structure. In the present chapter we discuss some features of the atomic structure of surfaces with particular emphasis on the possible departures from ideality. It is believed that surface defects play an extremely important role not only in atomic processes at surfaces such as diffusion, adsorption, and nucleation, but also in determining the electrical characteristics. Recent research on catalysis at Pt single crystal surfaces has indicated the importance of surface steps as active sites for the promotion of chemical reactions.

4.2. IDEAL SURFACES

Consider a perfect three-dimensional monatomic crystal. We may imagine the formation of a surface of orientation (hkl) by removing all atoms whose centers lie on one side of a mathematical plane of this orientation located within the crystal. If no relaxation or atomic rearrangement occurs we say that the resulting surface is *ideal*. [A useful collection of models of surfaces formed in this way can be found in the book by Nicholas (see bibliography).] Figure 4.1(a) and (b) shows models of ideal (100) and (111) surfaces of a face-centered cubic solid. Such surfaces are often referred to as being "atomically smooth" and are generally termed *singular* since it is believed that it is at orientations of this type that γ-plot singularities should occur. A surface which differs only slightly in orientation from one that is atomically smooth will consist mainly of flat portions or terraces with a system of widely spaced atomic *steps* or *ledges*. Such a surface is termed *vicinal*. Figure 4.2(a) shows an example of a vicinal surface which is at about 11° to the (100) plane and in which the steps are atomically straight. In general, however, the steps on a vicinal sur-

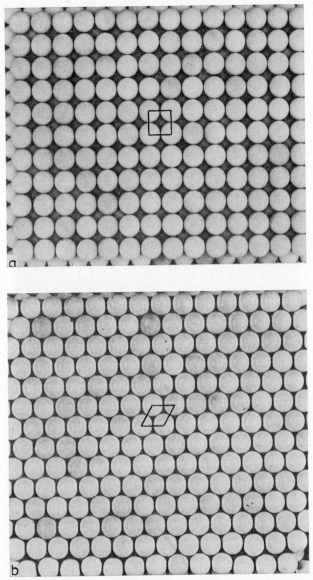

FIG. 4.1. Ball models of (a) an ideal (100) surface, and (b) an ideal (111) surface of a face-centered cubic monatomic crystal. The primitive unit meshes of the two-dimensional nets are indicated. Such surfaces are termed singular. (Courtesy of A. J. W. Moore.)

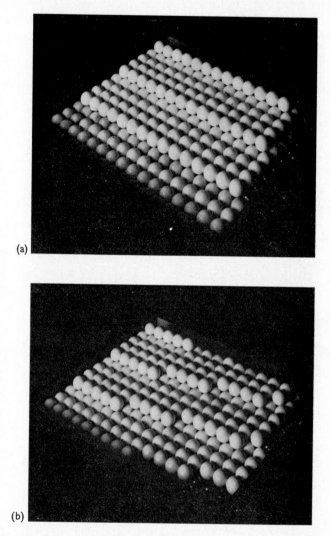

FIG. 4.2. (a) An ideal surface of a face-centered cubic crystal vicinal to the (100) plane showing straight monatomic steps. (b) An ideal surface vicinal to (100) with kinks along the steps. (Courtesy of A. J. W. Moore.)

face will not be completely straight, and *jogs* or *kinks* must be present to produce the desired orientation Fig. 4.2(b). The description of surfaces in terms of terraces, steps, and kinks was used extensively in the work of Kossel and Stranski. For a monatomic solid the kink site on a vicinal surface is of special interest since the removal of an atom from such a site to infinity regenerates another kink site and otherwise leaves the crystal unaltered. The work required for such a process thus corresponds to the binding energy per atom of the crystal. We shall later make use of this property of the kink site in discussing point defects near surfaces.

Evidence that such ideal surfaces can at least be approached in practice has been obtained from low energy electron diffraction and field ion microscopy studies. Figure 4.3 shows diffraction patterns from a singular and from a vicinal copper surface. The location of the diffraction spots is, to within experimental error, the same as that calculated on a model of regularly spaced monatomic steps as indicated in Fig. 4.3(c). Since the diffraction pattern reflects the average arrangement there may still be some departures from strict periodicity in the step distribution. Figure 4.4 is a field ion microscopy photograph (see section 6.2.5) of a specimen which is in the form of a sharp needle with an [001] axis. The surface is not an equilibrium one but has been produced artificially by the process of field evaporation. The positions of the "atoms" on the image are in good agreement with those predicted from a hard sphere model of the specimen surface.

It is possible to construct a theory to predict the shape of γ-plot cusps on the basis of a terrace and ledge model of vicinal surfaces. For a surface at a small angle θ to the singular plane which possesses an array of noninteracting linear monatomic steps of height h, the surface tension may be written as

$$\gamma = \gamma_s \cos \theta + \frac{\varepsilon}{h} \sin |\theta|, \qquad (4.1)$$

where γ_s is the surface tension of the singular surface and ε is the excess free energy associated with unit length of step. (This quantity is defined in a way quite analogous to that of surface free energy (section 1.2).) The effect of mutual step interactions may also be repre-

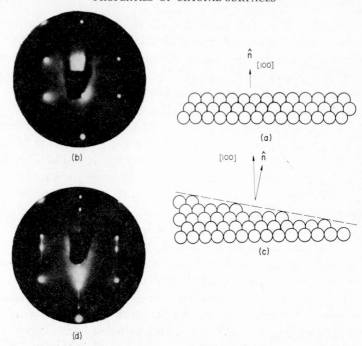

FIG. 4.3. (a) Side view of a model of an ideal (100) surface of a face-centered cubic crystal. (b) Back reflection electron diffraction pattern from a (100) copper surface. (c) Model of a surface at approximately 12° from (100) with an array of regularly spaced monatomic steps. (d) The electron diffraction pattern (126 eV) from a surface of a copper crystal cut at 12° to (100). Comparing (b) and (d) it will be seen that certain diffraction spots characteristic of the singular surface are replaced by pairs of spots for the vicinal surface. The spacing of the diffraction spot is inversely proportional to the average distance between the steps on the crystal surface. See section 6.3 for a discussion of low energy electron diffraction. (From G. E. Rhead and J. Perdereau in *Structure and Properties of Metal Surfaces*, CNRS, Paris, 1969.)

sented by including quadratic and higher order terms in the linear step density. Applying eqn. (4.1) to experimental data on the slopes at γ-plot cusps will yield an estimate of ε for different step directions. For example, it has been found from the surface tension data indicated in Fig. 3.9(a) that the excess free energy per unit length of monatomic step on the (100) surface of copper is about 1.8×10^{-6} ergs/cm

FIG. 4.4. Field ion micrograph of an iridium specimen taken with helium as the imaging gas. The arrangement of bright spots essentially corresponds to the positions of atoms on the approximately hemispherical tip although not all atoms in the exposed planes produce a bright spot image. The surface was prepared by the process of field evaporation (see section 6.2.5). The indices correspond to the directions of the normals to the portion of the surface indicated. (Courtesy of E. W. Müller.)

at 1100°K and about 0.5×10^{-6} ergs/cm at 1300°K, i.e. of the order of 0.01 eV per atom along the step. The temperature dependence of ε may be expressed as an excess entropy associated with the step; this will arise from differences in the vibrational frequencies of atoms at steps and also from the many possible configurations of the step array.

4.3. REAL SURFACES

4.3.1. *Surface Relaxation*

A surface which is formed at low temperatures, by cleavage, for example, may essentially retain the configurations described above but with small displacements or relaxations of the atoms or ions from the ideal surface positions as indicated in Fig. 4.5 for a monatomic crystal. It is to be expected that the fractional displacements δ_i

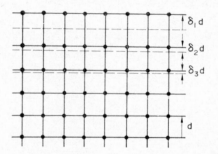

FIG. 4.5. Possible displacements of atoms near the surface of a crystal. The dotted lines indicate the positions of the planes at an ideal surface.

of the layers will decrease rapidly with depth into the crystal. A number of calculations have in fact been made of the relaxations to be expected for singular planes of various crystals, notably the alkali halides and inert gas crystal for which pair potentials are expected to give a fairly realistic representation. In most calculations it has been assumed that the displacements are in the direction of the surface normal although it has been suggested that on some alkali halide surfaces there may be lateral displacements to form rows of ion doublets. The case of ionic crystals is particularly interesting since the relaxation of the surface ions may produce an appreciable electrostatic dipole with a corresponding potential drop across the surface region. The calculated equilibrium positions near a (100) surface of NaCl is shown schematically in Fig. 4.6 from a model in which the top five layers were allowed to relax normal to the surface. The total dipole moment of such a surface would be[†] about 0.035 D per ion pair on the outermost plane, and this would produce a potential drop across the surface of about 85 mV.

A number of calculations have also been made on the relaxation of surface layers of metallic crystals using a model in which the binding is represented in terms of interactions between pairs of atoms. Although this is a convenient representation it must be regarded as being at best approximately correct particularly when it is used in calculating properties of a region in which the electron density is varying rapidly. Some results for copper obtained using a Morse interaction potential are shown in Table 4.1. The general trend of such results is that the surface layer is displaced by a few percent and the

TABLE 4.1. CALCULATED NORMAL DISPLACEMENTS OF
LAYERS NEAR FREE SURFACES OF COPPER

Surface	δ_1/d	δ_2/d	δ_3/d	δ_4/d
(100)	0.129	0.033	0.008	0.001
(110)	0.196	0.047	0.019	0.003
(111)	0.055	0.009	0.001	

δ_i is the displacement of the ith layer and d the interplanar spacing normal to the surface. These calculations were made using a Morse type of pairwise interaction potential. (From P. Wynblatt and N. A. Gjostein, A calculation of relaxation, migration and formation energies for surface defects in copper, *Surface Sci.* **12**, 109 (1968).)

[†] The unit of dipole moment is the Debye (D), equal to 10^{-18} e.s.u. (0.33×10^{-29} coul-m); it is a convenient unit for expressing the dipole moments of molecules since it is of the order of the electronic charge times an interatomic distance.

7*

88

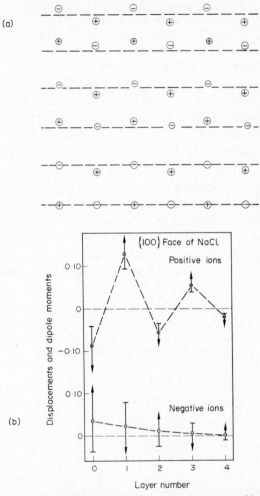

FIG. 4.6. (a) Schematic representation of the displacements of ions near the (100) surface of an alkali halide. The dotted lines indicate the positions of successive planes in the unrelaxed crystal. (b) Calculated ionic displacements and electronic dipole moments for cations and anions in the outermost (100) layers of a NaCl crystal. For these calculations the first five layers of the crystal were allowed to relax from their ideal crystal positions. The displacements are in units of the interplanar spacing with positive values indicating displacement in the direction of the outward normal. The direction and magnitude of the electronic dipole moments are indicated by the arrows, the unit of dipole moment being the Debye. The dipole moment of the anions

displacement decreases rapidly with layer number. It will also be noted that the smallest fractional displacements are predicted for the most closely packed surfaces. Recently estimates of the displacement of the outermost plane in a number of metals have been made in connection with surface energy calculations in which the ions were represented by pseudopotentials and the electron interactions treated in a self-consistent way. These indicate that the surface layer will be displaced outward with δ, ~ 0.005.

As yet there are no convincing experimental data on the question of surface relaxation although there are a number of observations in low energy electron diffraction which would be consistent with a distortion of the unit cell near the surface. As the details of the diffraction process become better understood we may expect quantitative information on surface relaxation to emerge.

4.3.2. Surface Point Defects

Apart from these relatively minor relaxations real surfaces may differ from the ideal configurations in ways which involve the displacement of atoms or ions over several interatomic distances. In fact most of the imperfections which exist in the bulk of crystals have their counterpart at the free surface. Figure 4.7 illustrates some of the defects which may exist on surfaces—vacancies and interstitials on terraces or in steps, self-adsorbed atoms on terraces and at steps, dislocation intersections with surfaces; there will also be groups of defects such as divacancies, steps which are multiples of interatomic distances in height, etc. There is good reason to believe that many of these defects will be involved in surface phenomena such as adsorption and condensation. Nucleation rates for condensation from the vapor are dependent on the surface geometry and can be influenced tremendously by the presence of surface steps and dislocation–surface intersections (Fig. 6.4).

---

is much greater due to their greater polarizability; for the anions the lengths of the arrows correspond to one-tenth of the dipole moment. (From G. C. Benson and K. S. Yun, *Advances in Chemistry Series* **33**, 26 (1961).)

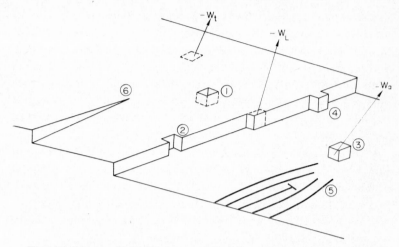

FIG. 4.7. Diagram illustrating some of the defects which may occur at a free surface. The examples shown are a vacancy on a terrace (1), a vacancy at a step (2), a self-adsorbed atom on a terrace (adatom) (3), a self-adsorbed atom at a ledge or step (4), an edge dislocation–surface intersection (5), and a screw dislocation–surface intersection (6). The diagram refers to a vicinal surface; with high index planes the classification is somewhat more difficult. The work to remove atoms from different types of sites on the surface is indicated. Note that in the removal of an atom from a kink site the kink site is regenerated.

Not all of the various defects mentioned above will be present on a crystal surface under equilibrium conditions but may nevertheless exist for kinetic reasons. It is fairly readily shown that dislocations are not equilibrium defects in crystals since the increase in configurational entropy is insufficient to offset the large energy increase. However, in a typical well-annealed single crystal the number of dislocation–surface intersections will be of the order of 10^4–10^6/cm^2.

At sufficiently high temperatures the concentrations of surface vacancies, interstitials, and self-adsorbed atoms should correspond to equilibrium, and expressions for these concentrations can then be obtained as follows. Let the work to remove to infinity an atom from a kink site, a complete terrace and an isolated adsorption site be $-W_L$, $-W_t$, and $-W_a$ respectively. (W_L is in fact the free energy of binding per atom of the perfect crystal.) These energies are shown schematically in Fig. 4.7. If M' is the number of adsorption sites per unit area and M the number of atoms per unit area in terrace sites, the numbers of self-adsorbed atoms n_a and vacancies n_v per unit area are given by

$$\frac{n_a}{M'} = \frac{1}{\exp\left(\dfrac{W_a - \mu}{kT}\right) + 1}$$

and

$$\frac{n_v}{M} = 1 - \frac{1}{\exp\left(\dfrac{W_t - \mu}{kT}\right) + 1}, \tag{4.2}$$

where μ is the chemical potential of atoms in the crystal and we have, of course, assumed that there is no defect–defect interaction. (We shall examine later some consequences of defect–defect interactions.) It is readily shown[†] from consideration of bulk properties only that μ is only slightly less than W_L and we obtain at sufficiently low temperatures

$$\frac{n_a}{M'} = \exp\left(-\frac{\Delta G_a}{kT}\right)$$

and

$$\frac{n_v}{M} = \exp\left(-\frac{\Delta G_v}{kT}\right). \tag{4.3}$$

$\Delta G_a\,[=(W_a - W_L)]$ and $\Delta G_v\,[=(W_L - W_t)]$ are called the free energies of formation of the self-adsorbed atom and the vacancy respectively; they can obviously be divided into internal energy and entropy contributions as is customary in discussions of bulk defects.

There have as yet been no direct experimental determinations of surface point defect concentrations or energies, so that at present we are forced to rely on theoretical estimates. Unfortunately the problem of calculating the energies is even more difficult than for the corresponding bulk case so that present results should be treated with some caution. The earliest estimates of the various energies in Fig. 4.7 were

---

[†] See, for example, the discussion by C. Herring in *The Physics of Powder Metallurgy* (ed. W. E. Kingston), McGraw-Hill, New York, 1951.

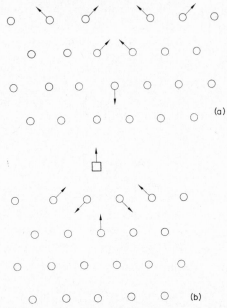

(a)

(b)

FIG. 4.8. Side views of the types of relaxation that occur around (a) a vacancy and (b) a self-adsorbed atom at a (100) argon surface. The arrows indicate the directions of displacement from the ideal crystal positions. (After J. J. Burton and G. Jura, The configuration and energy of defects on the (100) surface of a molecular crystal, in *Fundamentals of Gas–Surface Interactions*, ed. H. Saltsburg *et al.*, Academic Press, 1967, p. 75.)

apparently obtained using the nearest neighbor quasi-chemical model in which the energy per bond is considered to be fixed independent of the particular coordination of an atom. In such a scheme the entropy contributions are generally neglected and the bond energies are computed from the latent heat of sublimation ΔH_s per atom. For example, for a diamond cubic crystal for which the number of nearest neighbors is 4 the energy per bond ϕ is given by $\Delta H_s/2$, about 1.9 eV for germanium. This type of model is probably best for the homopolar crystals, not very appropriate for metals, and completely inappropriate for ionic crystals where interionic potentials are of very long range. For a (111) germanium surface we would obtain for the energy of formation of a vacancy (keeping the total number of atoms fixed) $3\phi - \Delta H_s = 1.9$ eV. Similar calculations are shown in Table 4.2 for particular surfaces of a few crystals. For the sake of comparison we may note that the nearest neighbor quasi-chemical model gives values of about 3.8 eV for the formation energies of bulk vacancies both in gold and in germanium, whereas the experimental results are about 1 eV and about 2.6 eV respectively.

Some improvement in the quasi-chemical model as applied to metals can be made by allowing the energy per bond to depend upon the number of bonds in which an atom is involved. Using experimental data on the heat of sublimation, surface tension, or bulk vacancy formation energies, it is readily shown by comparison with expressions for these quantities in terms of bond energies that the effective bond energies for metals increases significantly as the coordination decreases. Using this approach, the values given in Table 4.2 for gold are obtained. It will be seen that the values predicted are significantly less than those from the rigid bond models as we would expect. We should also note that if the energies are as low as these calculations indicate, the concentrations predicted from eqns. (4.3) are such that we can no longer neglect defect–defect interactions, and cooperative effects should become important well below the melting point. We shall consider this problem later. Calculations of surface defect energies using a pair potential of the Morse type (with normal bulk parameters) have also been made for different planes of copper, and the results are included in Table 4.2. Although this is probably somewhat better than the rigid bond model, it is not entirely realistic as previously mentioned.

A few calculations of surface defect energies have also been made for (100) surfaces of alkali halide crystals. Using a method of calculation similar to that used by Mott and Littleton[†] for bulk defect calculations, the value obtained for the energy to form a Schottky pair on the outermost (100) plane of NaCl was about 2.1 eV slightly greater than the corresponding bulk value of about 1.9 eV. However,

[†] See N. F. Mott and M. J. Littleton, *Trans. Faraday Soc.* **34**, 485 (1938).

TABLE 4.2. SOME ESTIMATES OF THE ENERGIES ASSOCIATED WITH SURFACE POINT DEFECTS

Surface	Vacancy formation energy (eV)	Adatom formation energy (eV)	Potential used
Cu (100)	1.17 (2ϕ)	1.17	Rigid nearest
(110)	0.58 (ϕ)	0.58	neighbor
(111)	1.75 (3ϕ)	1.75	model
Cu (100)	0.49	1.02	Morse[a]
(110)	0.51	0.60	potential
(111)	0.84	0.99	
Si (100)	0	0	Rigid nearest
(110)	2.32 (ϕ)	0	neighbor
(111)	2.32 (ϕ)	2.32 (ϕ)	model
Au (100)	1.26 (2ϕ)	1.26	Rigid nearest
(110)	0.63 (ϕ)	0.63	neighbor
(111)	1.89 (3ϕ)	1.89	model
Au (100)	0.25	0.57	Modified[b]
(110)	0.13	0.22	nearest neighbor
(111)	0.36	1.06	model
Ar (100)	0.033	0.021	Leonard–Jones[c] 6–12 potential
	0.026	0.026	Rigid nearest neighbor model[d]
NaCl (100)	2.12		Born–Mayer potential

[a] From P. Wynblatt and N. A. Gjostein, A calculation of relaxation, migration and formation energies for surface defects in copper, *Surface Sci.* **12**, 109 (1968).

[b] From R. L. Schwoebel, *J. Appl. Phys.* **38**, 3154 (1967); the values given here are obtained using a semi-empirical model in which the energy per bond is expressed as a quadratic function of the coordination number.

[c] From J. J. Burton and G. Jura in *Fundamentals of Gas–Surface Interactions* (ed. H. Saltsburg, J. N. Smith, and M. Rogers), Academic Press, New York, 1967. The pair potential used was

$$\phi_{ij} = (\beta/r_{ij}^{12} - \alpha/r_{ij}^{6}),$$

where r_{ij} is the pair separation and α and β are parameters determined from bulk properties. *(Continued on page 95)* the electronic and ionic polarization in the surface region due to the defects is very difficult to treat exactly, and further work is required on this particular problem.

The types of relaxation found for the atoms near a vacancy and an adsorbed atom at the (100) plane of argon are shown schematically in Fig. 4.8. The calculations for this case were carried out using a Leonard–Jones 6–12 pair potential which is known adequately ot describe a number of bulk properties, and energies were found for the formation of a vacancy in the topmost layer, a self-adsorbed atom of argon, the adsorption of neon and krypton and the presence of neon and krypton as substitutional impurities in the topmost layer. The result for the vacancy formation energy was about 0.033 eV and for the formation of the argon adatom about 0.021 eV. It is interesting to compare these with the result from a nearest neighbor rigid bond model which gives about 0.026 eV ($=2\phi$) for each defect. The difference is due to the greater binding of both the adatom and the terrace atom than that predicted from the nearest neighbor model.

4.4. DEFECTS NEAR IONIC CRYSTAL SURFACES

For homopolar and metallic crystals for which the effective interatomic or inter-ionic interactions are of short range, we do not expect the point defect concentration to be disturbed for more than a few atomic layers below the free surface. However, for ionic solids the inter-ionic potential is of very long range and the defect distribution may be disturbed for appreciable distances into the crystal. In these crystals the vacancies and interstitials are effectively charged, and in the interior of the crystal their concentrations will be such as to

---

[d] From J. M. Blakely and C. Y. Li, *J. Phys. Chem. Solids* **26**, 1863 (1965). The pair potential used was

$$\phi_{ij} = \pm\frac{e^2}{r_{ij}} + A \exp(-r_{ij}/\varrho)$$

where A and ϱ are determined from bulk properties. The energy quoted is for the formation of a pair of vacancies on the topmost plane.

preserve charge neutrality. However, near the surface (or other imperfection such as a dislocation or grain boundary) the concentrations may be changed resulting in the formation of a space charge region. Such space charge regions near lattice discontinuities in ionic crystals were postulated by Frenkel and have been discussed by a number of authors for a variety of conditions. The following is one way of formulating the problem in which some account is taken of the surface properties. We will consider an ionic crystal, such as AgCl or AgBr, in which the bulk point defects consist of cation interstitials and cation vacancies with the anion sublattice remaining essentially perfect. In the bulk of the crystal interstitials and vacancies are created simultaneously so that if n_I and n_v are the numbers of interstitials and vacancies per unit volume respectively, $n_v = n_I$ and

$$n_I = \sqrt{(NN_I)} \exp\left(-\frac{\Delta G_F}{2kT}\right), \tag{4.4}$$

where N and N_I are the numbers of lattice and interstitial sites per unit volume respectively and ΔG_F is the free energy associated with the formation of a Frenkel pair (see Fig. 4.9(a), p. 97). The chemical potential μ_{Ag} of the silver ions in the crystal may be obtained from the charge neutrality condition giving

$$\mu_{Ag} = W_+ + \frac{1}{2}\Delta G_F - \frac{1}{2}kT \ln\left(\frac{N_I}{N}\right) \tag{4.5}$$

and is also indicated in Fig. 4.9. W_+ is the work to remove a cation from a normal lattice site leaving a vacancy.

When a free surface or other lattice imperfection is present it becomes possible for the vacancies and interstitials to be created and annihilated separately. We will suppose for simplicity that the exchange of positive ions between the top surface layer and the crystal interior involves a particular type of surface site, e.g. the kink sites, where the binding energy of the positive ion is W_+^s indicated schematically in Fig. 4.9(b).[†] For each surface orientation, and with a perfect

---

[†] The energy levels indicated in Fig. 4.9 are to be regarded as effective "one-ion" energies. They represent the relative energies of the crytal with the ion in different possible sites.

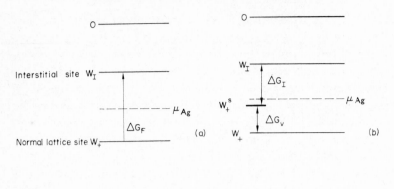

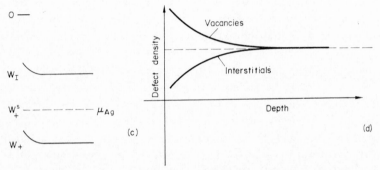

FIG. 4.9. (a) Schematic arrangement of various energies associated with cations in an ionic crystal. The zero of reference is that of an isolated cation at rest at infinite distance from the crystal. W_+ is the work to remove a cation to infinity from a normal site leaving a vacancy, W_I is the work to remove a cation from an interstitial site to infinity, μ is the electrochemical potential of cations, and ΔG_F is the free energy of Frenkel pair formation. In (b) the energy to remove a cation from a particular surface site has been introduced. For any particular surface site one can define the free energies ΔG_V and ΔG_I involved in forming a cation vacancy and an interstitial respectively. In (c) the occupation of the empty surface sites has caused μ to coincide with W_+^s and has generated an electrostatic potential difference between the surface and interior of the crystal. The defect distribution corresponding to this situation is shown schematically in (d).

chlorine sublattice, the number N^s of such surface kink sites per unit area will be fixed (neglecting thermal creation of kink sites), and when the surface is uncharged half will be occupied by silver ions and half occupied by chlorine ions. The position in energy of the surface site relative to the chemical potential of the silver ions will, of course, determine its occupancy. If, for example, W^s lies below μ, the surface will become positively charged. There will be a corresponding negative space charge region with accompanying "bending" of the energy levels for the various types of sites as the surface is approached. The problem is formally very similar to that of an intrinsic semiconductor with a single set of surface electronic states at fixed energy (section 5.5). With a sufficiently large density of surface sites at energy W^s_+ the occupation of the sites will maintain the chemical potential nearly coincident with W^s_+, Fig. 4.9(c), and the distribution of defects in the crystal will be as shown schematically in Fig. 4.9(d); at high temperatures the density of defects in the crystal may become sufficiently large that the surface level cannot accommodate enough silver ions to maintain the chemical potential at W^s_+, and the magnitude of the potential across the space charge region will be decreased.

For any particular surface it is clear that we can define values of the separate free energies of formation of the vacancy and the interstitial, ΔG_v and ΔG_I, as indicated in Fig. 4.9(b). These may be expected to vary with the orientation of the surface so that different surfaces of the same crystal may be at different electrostatic potentials and there will therefore be electrostatic fields outside the surfaces of small crystals. The difference in electrostatic potential between surface and bulk may be several hundred millivolts with a penetration depth of the space charge region of several hundred angstroms.

Generally, in an ionic crystal such as AgCl there will be a sufficient concentration of divalent cation impurities to have a significant effect upon the silver ion chemical potential and defect concentrations, and in such cases it can be shown that the chemical potential will vary more strongly with temperature than in pure or intrinsic crystals. It will be possible that at some temperature the chemical potential of silver ions will coincide with the energy of the surface site: in such circumstances there will be no surface charge and no space charge region, and we refer to this as an *isoelectric temperature*.

There is a fairly substantial amount of experimental evidence that such space charge regions exist near lattice discontinuities in ionic crystals, and in cases where electronic effects can be neglected quantitative studies are expected to yield information on the energies and densities of different surface sites. With the silver halides the ionic space charge region is likely to be of importance in connection with the formation of latent images due to the influence of the electrostatic field on the trajectories of photo-created electrons and holes. Thus, for example, if the surface is negatively charged at equilibrium it will attract holes and repel electrons. The latter may then combine with interstitial silver ions to produce the photographic latent image. The space charge region should also be of importance in determining the surface conductivity and sintering characteristics of many ionic crystals, and may be of importance in the growth of oxide films.

4.5. COOPERATIVE EFFECTS
AMONG SURFACE DEFECTS

In section 4.3.2 we have assumed in writing down the surface defect concentrations that the numbers of defects were so small that all defect–defect interactions could be neglected. However, we have seen that the estimates for the energies of formation could be rather small, and this would result in large concentrations at temperatures well below the normal melting point. For example, if we assume a value of about 0.25 eV for the energy of formation of a vacancy on the (100) plane of gold, the corresponding fractional concentration of vacancies assuming no mutual interaction would be about 10^{-1} at 1000°K. In such cases, due to mutual interactions, the energies of formation become concentration dependent, and this may lead to catastrophic increases in defect concentrations with increase of temperature. Such a process is generally referred to as *surface roughening* or *surface melting*. It is clear that if this occurred below the melting point it would have a strong influence on many surface phenomena associated with crystals. Quantitative predictions on the temperature range in which surface roughening is likely to occur requires a knowledge of a number of energies which are not yet established. These are the energies associated with the formation of surface vacancies, adatoms,

steps, kinks, and various combinations of these defects. Some qualitative aspects of the effect can, however, be illustrated by considering
a very simple case. We consider the roughening of a singular surface
of a one-component crystal by the formation of vacancies and adatoms
allowing for the possibility that pairs of these individual defects may
associate to form divacancies and di-adatoms. In the interest of
minimizing the number of symbols we shall consider the particular
case of the roughening of a (100) surface of a face-centered cubic
crystal.[†] On forming n_v vacancies, n_a adatoms,[‡] n_{2v} divacancies, n_{2a}
di-adatoms per unit area on which there are M atom sites per unit
area, the change in Gibbs free energy per unit area of surface of a
semi-infinite crystal may be written

$$\delta G = (n_v \, \Delta G_v + n_{2v} \, \Delta G_{2v} - kT \ln \omega_v) + (n_a \, \Delta G_a + n_{2a} \, \Delta G_{2a} - kT \ln \omega_a),$$
$$(4.6)$$

where ΔG_v and ΔG_a are defined in section 4.2.2, ΔG_{2v} is the increase
in free energy accompanying the coalescence of a pair of isolated sur-
face vacancies and ΔG_{2a} the corresponding quantity for adatoms.
ω_v and ω_a are the numbers of ways of distributing the vacancies and
adatoms among the available sites. The number of divacancies per
unit area can be expressed approximately as

$$n_{2v} = 4 \frac{n_v^2}{M} \exp\left(-\frac{\Delta G_{2v}}{kT}\right) \qquad (4.7)$$

with a similar expression for n_{2a}. By noting that the formation of a
vacancy effectively excludes four sites for adsorbed atoms and that
a divacancy excludes six adatom sites,

$$\omega_a = \frac{(M - 4n_v + 2n_{2v})!}{(M - 4n_v + 2n_{2v} - n_a)! \, n_a!}, \qquad (4.8)$$

[†] The treatment given here is similar to that in J. P. Hirth, *Energetics in Metallurgical Phenomena*, Vol. 2 (ed. W. M. Mueller), Gordon & Breach, 1965. A more
complete treatment of the phenomenon is given in the paper by Burton *et al.*, listed
in the bibliography.

[‡] n_v and n_a are the *total* numbers of vacancies and adatoms respectively.

with a similar expression for ω_v. Assuming that adsorbed atoms and vacancies can be formed independently on the singular surface (e.g. by exchange of atoms between the surface and bulk defects) the equilibrium concentrations of the defects are found from the conditions

$$\frac{\partial}{\partial n_a}(\delta G) = \frac{\partial}{\partial n_v}(\delta G) = 0 \tag{4.9}$$

at constant temperature and total number of atoms. Performing the algebra for the (100) surface gives

$$\frac{n_v}{(M-4n_a+2n_{2a}-n_v)} = \exp\left[-\frac{\Delta G_v}{kT}\left\{1+8\,\frac{n_v}{M}\,\frac{\Delta G_{2v}}{\Delta G_v}\exp\left(-\frac{\Delta G_{2v}}{kT}\right)\right\}\right] \tag{4.10}$$

and a corresponding expression for n_a. The important feature to note in comparing this expression with that in eqn. (4.3) is that when the concentrations of the defects are sufficiently large the average energy associated with the formation of each defect becomes concentration dependent. If ΔG_{2v} is negative, i.e. the free energy is reduced by joining a pair of isolated vacancies, then eqn. (4.10) shows that the effective vacancy formation energy is decreased with increasing temperature and vacancy concentration. If we use a nearest neighbor rigid bond model in which the energy per bond is ϕ and neglect vibrational entropy contributions to the defect free energies, it is readily shown that eqn. (4.10) becomes

$$\frac{x_v}{(1-4x_a+2x_{2a}-x_v)} = \exp\left[-\frac{2\phi}{kT}\left\{1-4x_v\exp\left(-\frac{\phi}{kT}\right)\right\}\right], \tag{4.11}$$

where x_v, x_a, x_{2a} are the fractions of the available sites which are vacant, are occupied by an adsorbed atom or have a di-adatom associated with them. There is, of course, an analogous expression for adatoms. A convenient way of expressing the degree of surface roughness is as the average number S per surface site of free bonds parallel to the surface. Using eqns. (4.7) and (4.11) we obtain in the nearest neighbor rigid bond model

$$S = 8x\left\{1 - x \exp\left(\frac{\phi}{kT}\right)\right\}, \qquad (4.12)$$

where $x = x_v = x_a$. The variation of S with temperature is shown schematically in Fig. 4.10 as a plot of S versus kT/ϕ. There is little

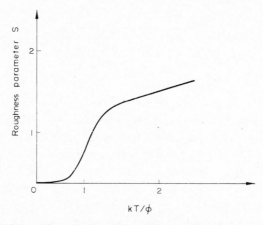

FIG. 4.10. Schematic variation of the surface roughness parameter S with the quantity kT/ϕ as predicted by a nearest neighbor type of interaction model. ϕ is the energy per bond between a pair of neighboring atoms. For $kT/\phi > 1$ the number of missing neighbors per surface atom approaches unity. (After J. P. Hirth in *Energetics in Metallurgical Phenomena*, ed. W. M. Mueller, Gordon & Breach, 1965.)

roughening at temperatures for which $kT/\phi \lesssim 0.5$; above this temperature it increases very rapidly. For $\phi \approx 1$ eV substantial roughening should be present only above about 6000°K, considerably higher than the melting point. However, in view of the crudeness of the nearest neighbor type of model, particularly for crystals other than strongly covalently bonded ones, the possibility of substantial surface roughening should not be discounted. An effect of this would be to diminish the orientation dependence of surface properties with increasing temperature since singular surfaces should lose their unique character and resemble ordinary high index surfaces.

It should be emphasized that we have considered here only a very special case of surface roughening due to the spontaneous creation of point defects. Clearly the thermal creation of surface steps may also occur as well as the roughening of steps on vicinal surfaces by the formation of kinks. The latter would allow the steps to wander over the surface, an effect which greatly increases the configurational entropy associated with the step array. Calculations based on a nearest neighbor model do, in fact, indicate that kinks should be created in significant densities on steps at temperatures well below the melting point.

4.6. INTERACTIONS BETWEEN SURFACE STEPS

There are a number of experimental observations on crystal surfaces which suggest that there is some form of interaction between steps. Low energy electron diffraction observations of annealed vicinal surfaces of copper (see Fig. 4.3) and of UO_2 indicate that the steps assume a fairly regular array consistent with a repulsive type of interaction between like steps. There are also a number of experiments in which sinusoidal surfaces (which can be made up of arrays of steps with a sinusoidal variation in spacing) are observed to decay in amplitude at high temperatures (see Fig. 7.6). These latter observations appear to be consistent with repulsive forces between like steps and attractive forces between unlike steps, and it should, in fact, be possible to provide a description of mass transport processes at surface (see section 7.3) in terms of the interactions between surface steps. The origin of these interactions is not yet clear. As indicated in the previous section, it is probable that at temperatures below the melting point kinks will be spontaneously created on surface steps and there will be a tendency for wandering of steps in order to increase configurational entropy. The coalescence of either a pair of like or a pair of unlike steps (Fig. 4.11) should lead to a net decrease in

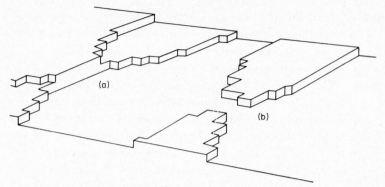

FIG. 4.11. Interaction of steps on a vicinal surface. At (a) the like steps have coalesced over a part of their length to form a section of step which is double that of the individual steps. At (b) partial annihilation of unlike steps has occurred.

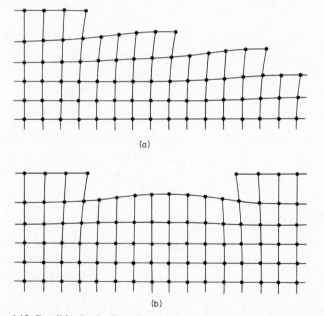

FIG. 4.12. Possible elastic distortions associated with pairs of (a) like steps and (b) unlike steps on a crystal surface. (From J. M. Blakely and R. I.. Schwoebel, *Surface Sci.* **26,** 321 (1971).)

energy; in the first case the energy decrease should be relatively small, but unlike steps will annihilate on coalescence, and hence in this case there should be a relatively large energy decrease. In both cases the configurational entropy is decreased and it is therefore possible that in some range of temperatures the entropy decrease will be large enough to result in a repulsive interaction for the like steps while that between the unlike steps will be attractive. Another source of the apparent step interaction may be associated with the normal displacements of atoms in planes near the surface, discussed in section 4.3.1. Suppose that the atoms are displaced from their ideal crystal positions in the direction of the surface normal by an amount which decreases with increasing depth. When a step is present on an otherwise atomically smooth surface, the displacements of the atoms will depend on their positions relative to the step so that a region of extra elastic strain may be associated with the step in a way that is somewhat analogous to the strain field associated with an edge dislocation in a crystal. Figure 4.12(a) and (b) indicate schematically the situation that may arise with pairs of like or unlike steps whose strain fields overlap. By analogy with edge dislocation interactions one would predict repulsive interactions in (a) and an attractive interaction in (b). However, the situation has not yet been analyzed quantitatively.

BIBLIOGRAPHY

BLAKELY, J. M., and LI, C. Y., Formation energies of vacancies at a (100) sodium chloride surface, *J. Phys. Chem. Solids* **26,** 1863 (1965).

BURTON, J. J., and JURA, G., The configuration and energy of defects on the (100) surface of a molecular crystal, in *Fundamentals of Gas–Surface Interactions* (ed. H. Saltsburg, J. M. Smith, and M. Rogers), Academic Press, New York 1967, p. 75.

BURTON, W. K., CABRERA, N., and FRANK, F. C., The growth of crystals and the equilibrium structure of their surfaces, *Phil. Trans. Roy. Soc.* A**243,** 299 (1951).

CABRERA, J., The structure of crystal surfaces, *Disc. of Faraday Soc.* **28,** 16 (1959).

DUNNING, W. J., The structure of surfaces and its role in adsorption, in *The Solid–Gas Interface*, Vol. 1 (ed. E. A. Flood), Dekker, New York, 1967, chap. 9.

ELLIS, W. P., and SCHWOEBEL, R. L., LEED from surface steps on UO_2 single crystals, *Surface Sci.* **11,** 82 (1968).

FRENKEL, J., *Kinetic Theory of Liquids*, Dover Publications, 1955.

GRIMLEY, T. B., The contact between a solid and an electrolyte, *Proc. Roy. Soc.* **A201**, 40 (1950).

GRUBER, E. E., and MULLINS, W. W., On the theory of anisotropy of crystalline surface tension, *J. Phys. Chem. Solids* **28**, 875 (1967).

HIRTH, J. P., The kinetic and thermodynamic properties of surfaces, in *Energetics in Metallurgical Phenomena*, Vol. 2 (ed. W. M. Mueller), Gordon & Breach, 1965.

KNACKE, O., and STRANSKI, I. N., The mechanism of evaporation, *Progress in Metal Physics* **6**, 181 (1956).

NICHOLAS, J. F., *An Atlas of Models of Crystal Surfaces*, Gordon & Breach, 1965.

POEPPEL, R. B., and BLAKELY, J. M., Origin of equilibrium space charge potentials in ionic crystals, *Surface Sci.* **15**, 507 (1969).

RHEAD, G. E., and PERDEREAU, J., LEED studies of vicinal surfaces, in *Structure and Properties of Solid Surfaces*, CNRS, Paris, 1969.

RHODIN, T., PALMBERG, P. W., and TODD, C. J., Surface point defects and epitaxial growth on alkali halides, in *Molecular Processes on Solid Surfaces* (ed. E. Drauglis, R. D. Gretz, and R. I. Jaffee), McGraw-Hill, New York, 1969.

SCHWOEBEL, R. L., Surface vacancies on metal crystals, *J. Appl. Phys.* **38**, 3154 (1967).

SOMORJAI, G. A., The surface structure of and catalysis by platinum single crystal surfaces, *Catalysis Reviews* **7**, 87 (1972).

WYNBLATT, P., and GJOSTEIN, N. A., A calculation of migration energies and binding energies for tungsten adatoms on tungsten surfaces, *Surface Sci.* **22**, 125 (1970).

CHAPTER 5

SOME THEORETICAL ASPECTS
OF SURFACE STUDIES

5.1. INTRODUCTION

A detailed understanding of surface phenomena will require a knowledge of the equilibrium structure and properties of the pure surface itself. By this we mean the time average coordinates of the atoms or ions in the surface region, the electronic states associated with surface atoms, and their dynamical behavior. In the present chapter we describe some of the theoretical work in this direction. Because the inert gas crystals and ionic crystals can be adequately represented, for many purposes, by pair potentials, the most detailed calculations of structure have been carried out for these cases. We will consider the method of calculating lattice relaxation near simple surfaces of these materials and the corresponding effects on surface tension. With metals the distribution of conduction electrons is expected to be perturbed appreciably at the surface unit cells. We will consider some aspects of the electronic charge distribution and surface tension of metals, and investigate the origin of the electronic work function and inner potential. Finally, we mention the existence of surface electronic states and surface phonons. Surface electronic states strongly influence the electrical properties of semiconductor surfaces where they lead to the formation of regions of electronic space charge extending into the crystal.

5.2. SURFACE STRUCTURE AND SURFACE TENSION OF IONIC CRYSTALS AND INERT GAS CRYSTALS

The calculation of the surface tension of a crystal is, of course, intimately connected with that of determining the equilibrium structure in the surface region. For ionic and inert gas crystals it is generally believed that many bulk properties can be adequately described by models based on pair potentials, and the same assumption is generally made in making surface computations on these materials. For this reason the number of theoretical papers on the surface energy of these crystals far exceeds that on other types of solids. These are critically reviewed in the article by Benson and Yun listed in the bibliography, and we shall outline some of their conclusions here.

From section 1.2 the surface energy e^s, surface entropy s^s, and surface tension are related by

$$\gamma(T) = e^s(T) - Ts^s(T). \qquad (5.1)$$

$e^s(T)$ will consist of changes in potential energy of interaction and vibrational energy and $s^s(T)$ will mainly be due to vibrations. Surface point defects and steps will also contribute to these quantities, but these contributions have generally been neglected in the theoretical work. We may express $e^s(T)$ in terms of the potential energy contribution at $0°K$, $e^s_{pot}(0)$, and the vibrational contribution $e^s_{vib}(T)$ as

$$e^s(T) = e^s_{pot}(0) + \int_0^T \left(\frac{\partial e^s_{pot}}{\partial T} \right) dT + e^s_{vib}(T), \qquad (5.2)$$

and hence

$$e^s(T) = e^s_{pot}(0) + \left(\frac{\partial e^s_{pot}(0)}{\partial a} \right) \varDelta a + e^s_{vib}(T), \qquad (5.3)$$

where $\varDelta a$ is the change in lattice parameter between $0°K$ and T. The potential energy contribution to the surface energy at $0°K$ is the quantity which has been the subject of most of the theoretical work. This contribution can be visualized as due to the breaking of bonds when a new surface is created. We can imagine the new surfaces to be created by first splitting the crystal into two parts with the atoms or ions in each half maintaining their relative spacings unchanged. This will require an amount of energy $e^{s0}_{pot}(0)$ per unit area of each surface. The atoms or ions are then allowed to relax to their new equilibrium position. This second step will involve an energy increase $\Delta e^{s}_{pot}(0)$. Thus the calculation of the surface tension involves evaluating the various terms in the formula

$$\gamma(T) = e^{s0}_{pot}(0) + \Delta e^{s}_{pot}(0) + \left(\frac{\partial e^{s}_{pot}(0)}{\partial a}\right)\Delta a + e^{s}_{vib}(T) - Ts^{s}_{vib}(T). \quad (5.4)$$

The evaluation of the potential energy terms might at first sight appear to be relatively straightforward (although by no means simple due to the lattice summations which are involved). However, there may be considerable difficulty in choosing the correct form of the pair interaction energy which is appropriate for atoms or ions in the surface region. This is particularly true for the ionic crystals where there is significant displacement and electronic polarization.

For a surface of one of the *inert gas crystals*, which are face-centered cubic, the term $e^{s0}_{pot}(0)$ is given by

$$e^{s0}_{pot}(0) = n \sum_{\lambda=0}^{\infty} (\varepsilon^{0}_{\lambda} - \varepsilon_{\infty}), \quad (5.5)$$

where n is the number of atoms per unit area in each successive plane parallel to the surface, $\varepsilon^{0}_{\lambda}$ is the potential energy of interaction between an atom in layer λ with all other atoms in the semi-infinite crystal before relaxation, and ε_{∞} is the corresponding quantity for an atom in the infinite crystal. For any atom i in layer λ the interaction energy is given in terms of the pair interaction potential ϕ_{ij} by

$$\varepsilon^{0}_{\lambda} = \tfrac{1}{2} \sum_{j} \phi_{ij},$$

the sum being taken over all other atoms of the crystal. Similarly, the change in potential energy due to relaxation is

$$\Delta e_{\text{pot}}^s(0) = n \sum_{\lambda=0}^{\infty} (\varepsilon_\lambda - \varepsilon_\lambda^0), \qquad (5.6)$$

where ε_λ represents the interaction energy for an atom in layer λ in the relaxed semi-infinite crystal. If d is the separation of layers which are parallel to the surface in the ideal crystal, the displacement of the layer λ in the relaxed state is taken as $Z_\lambda d$ in the direction of the outward normal, and the equilibrium positions are then determined from the set of simultaneous equations

$$\frac{\partial \Delta e_{\text{pot}}^s(0)}{\partial Z_\lambda} = 0 \qquad \lambda = 0, 1, \ldots (M-1) \qquad (5.7)$$

where M is the total number of layers which are allowed to relax. Computations of these displacements were carried out by Shuttleworth and more recently also by Benson and co-workers using both a Leonard–Jones potential and an exponential-6 potential. For both types of potential the surface planes were found to be displaced in the outward direction by an amount which decreases rapidly with depth. The values of the displacement were less than 5% of the interlayer spacing, the fractional displacement increasing in the order (111), (100) and (110).

With the *alkali halides* the problem is considerably more complex due to the two types of ions present and the more complicated interionic potential. For each (100) type of layer there are two ionic displacements and two electronic dipole moments, so that if a calculation is done in which M layers are allowed to relax, there are $4M$ unknowns with a corresponding number of simultaneous equations to be solved. Some results for surface displacements at a (100) surface of NaCl were given in Fig. 4.6(b), but it should be noted that the results are very sensitive to the form of the potential assumed and to the values of the input parameters.

The computation of the vibrational energy and entropy terms in eqn. (5.4) of course requires a knowledge of the distribution of normal modes of vibration with and without a surface. For any particular mode of vibration of frequency v the average energy $\varepsilon(v)$ is given by

$$\varepsilon(v) = \frac{1}{2}\,hv + \frac{hv}{e^{hv/kT}+1}$$

and by the definition of surface excess quantities (section 1.2) where we imagine the dividing surface to be located just outside the last layer of ions or atoms

$$e_{\text{vib}}^{s}(T) = \frac{1}{A}\int_{0}^{\infty} \varepsilon(v)\{D'(v) - D(v)\}\,dv, \qquad (5.8)$$

where $D'(v)\,dv$ and $D(v)\,dv$ are respectively the number of modes between v and $v+dv$ in the real crystal with a surface of area A and in the hypothetical case of a crystal containing the same number of atoms but with periodic boundary conditions applied at its surface. The vibrational contribution to the surface entropy will also be given by an equation of the same form as eqn. (5.8). In the Debye model of the lattice vibrations the frequency spectrum for an isotropic elastic continuum is used to approximate that of the crystal, i.e.

$$D(v) = \frac{12\pi V v^{2}}{c^{3}} \quad \text{for} \quad v \leqslant v_{D}, \qquad (5.9)$$

and $D(v)$ is zero for $v > v_{D}$, where the maximum or Debye frequency v_{D} is obtained from

$$\int_{0}^{v_{D}} D(v)\,dv = 3N, \qquad (5.10)$$

where N is the number of atoms in the crystal. It has been shown,[†]

See R. Stratton, *Phil. Mag.* **44**, 519 (1953); *J. Chem. Phys.* **37**, 2972 (1962).

using a continuum model, that for a finite solid with surface area A the normal Debye spectrum should be modified to

$$D'(v) = \frac{12\pi V v^2}{C^3} + \beta A \frac{v}{c^2} \quad \text{for} \quad v \leqslant v'_D, \qquad (5.11)$$

where A is the surface area and β is a function of the elastic constants of the medium, the new cutoff frequency v'_D again being obtained by equating the total number of modes to $3N$. Equations (5.8), (5.9), and (5.11) have been used to estimate the vibrational energy and entropy contributions in eqn. (5.4) for a number of cases, and some examples of the results are shown in Table 5.1. Although the modified

TABLE 5.1. EXAMPLES OF CALCULATED CONTRIBUTIONS TO THE SURFACE TENSION IN INERT GAS AND IONIC CRYSTALS

(Energies in ergs/cm²)

Crystal	Temperature (°K)	$e^{s_0}_{pot}(0)$	$\Delta e^s_{pot}(0)$	$\left(\dfrac{\partial e^s_{pot}(0)}{\partial a}\right)\Delta a$	$e^s_{vib}(T)$	$-Ts^s_{vib}(T)$	$\gamma(T)$
Argon[a] (111)	83.8 (melting point)	43.31	−0.14	−5.6	−0.47	−6.2	30.9
KCl[b] (100)	298	175.3	−34.0	4	−1.3	−25.6	118

[a] Calculated using a Leonard–Jones 6–12 potential (see p. 94).
[b] Calculated using a Born–Mayer–Huggins potential.
(From calculations given by G. C. Benson and K. S. Yun, Surface energy and surface tension of crystalline solids, in *The Solid–Gas Interface*, Vol. 1 (ed. E. A. Flood), Dekker, New York, 1967, chapter 8.)

Debye spectrum (5.11) can only be taken as an approximate model, the numbers in the table probably indicate the relative importance of different contributions to γ. The Debye model as outlined above is really only applicable to a monatomic solid, but it is probably also reasonable to use it for KCl where the masses of the two ions are essentially the same. For KCl both the relaxation energy and the entropy term are very significant at room temperature. In argon the entropy term has a magnitude of about 20% of the final surface tension while the lattice relaxation energy is insignificant. The difference in the importance of relaxation in the two cases is due to the long range potential in the alkali halide and the polarization of the surface region.

Unfortunately, sufficiently reliable experimental data with which to compare these computations are not yet available. Some low energy electron diffraction data on the (100) surface of LiF[†] show diffraction maxima which on a kinematic basis (see section 6.3) would be disallowed for a face-centered cubic lattice and may be interpreted in terms of distortion of the unit cell near the surface. However, dynamical effects in low energy electron diffraction are known to be very important, so that the interpretation of the observations is not yet entirely clear. Recently crystals of xenon and argon have been prepared for examination by low energy electron diffraction and with xenon in particular the interpretation of data on the diffraction of low energy electrons seems to be considerably simpler than with most other materials. However, the relatively small fractional expansions predicted for the surface spacing in this case is probably within the normal experimental error of most low energy electron diffraction measurements at present.

5.3. SEMI-EMPIRICAL ESTIMATES OF SURFACE TENSION IN CRYSTALS

The previous section has given an example of the calculation of surface tension from first principles. Estimates of γ can also be obtained from a number of semi-empirical relations in which the surface tension is related to other measured properties of the crystal such as the elastic constants, the heat of sublimation, the liquid surface tension, the heat of fusion, etc. We will consider two examples in the present section.

[†] See E. G. McRae and C. W. Caldwell, Low energy electron diffraction study of lithium fluoride (100) surface, *Surface Sci.* **2**, 509 (1964).

Suppose that when a crystal is split on a particular plane the potential energy of interaction per unit area of surface between the two half crystals $U(x)$ may be represented as a function of their separation x as in Fig. 5.1, curve (a). If we approximate the potential energy curve as being harmonic between d_0 and some distance d_0+a, curve

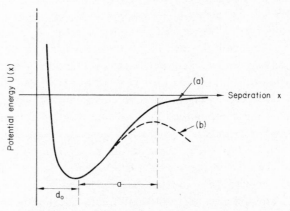

FIG. 5.1. Potential energy of interaction $U(x)$ between two half crystals as a function of the separation of the atoms in the outermost planes. d_0 is the equilibrium interplanar spacing. Curve (b) (dashed) is a harmonic approximation to the real curve (a) obtained by fitting at $x = d_0$.

(b), where d_0 is the equilibrium interplanar spacing normal to the surface, then the force per unit area between the two half crystals is

$$F(x') = F_0 \sin \frac{\pi x'}{a}, \qquad (5.12)$$

where $x' = (x-d_0)$, and the work done in moving the crystals apart becomes

$$\int_0^a F(x')\, dx' = F_0 \frac{2a}{\pi}. \qquad (5.13)$$

F_0 may be related to the Young's modulus E by fitting eqn. (5.12) to

Hooke's law at small displacements, i.e.

$$F_0 = E \frac{a}{\pi \, d_0}, \qquad (5.14)$$

so that the energy change in separation (when two surfaces are created) is

$$\frac{2E}{d_0} \left(\frac{a}{\pi} \right)^2. \qquad (5.15)$$

Formula (5.15) has been used quite frequently to obtain estimates of surface tension in crystals, and Table 3.1 compares some values calculated in this way with experimental results obtained by the cleavage technique. There is, of course, some question as to what value should be used for the parameter a and it should be noted that the results are very strongly dependent on this choice. It appears that best agreement is obtained by taking a to be the mean atomic or ionic radius of atoms or ions in the surface plane. With this choice, however, the agreement with the existing experimental data is rather remarkably good over a wide range of different materials.

Another frequently used semi-empirical scheme for estimating surface tensions is one which is based on the use of the quasi-chemical model discussed in section 4.3.2. The surface tension is estimated from the formula

$$\gamma = \tfrac{1}{2} \sum_{ij} N_j n_i^j \phi_j, \qquad (5.16)$$

where N_j is the number of atoms per unit area of type j (distinguished by its coordination number at the surface) and n_i^j is the number of bonds broken of type i for each surface atom of type j on forming the surface. Generally this formula is used by cutting off at nearest neighbor interactions although it has been extended to higher order interactions in a few cases. With nearest neighbor interactions only there is only one bond energy involved, and this is generally estimated from the latent heat of sublimation ΔH_s per atom as $(2 \, \Delta H_s)/z$, where z is the bulk coordination number. Evaluating the bond energy in this way would appear to neglect all entropy and lattice relaxation contri-

TABLE 5.2. ESTIMATES OF SURFACE TENSION
FOR SOME COVALENTLY BONDED CRYSTALS

Crystal	Surface tension (ergs/cm²)		
	{100}	{110}	{111}
C (diamond)	3064	2170	1770
Si	2513	1781	1451
Ge	1927	1365	1113
InSb	1100	750	600
GaAs	2200	1500	1300
InAs	1400	1000	840
GaSb	1600	1100	910
InP	1900	1300	1100
AlSb	1900	1300	1100
AlAs	2600	1800	1500
GaP	2900	2000	1700
AlP	3400	2400	2000

The experimental values for the {111}
planes of germanium and silicon are 1060 and
1230 ergs/cm² respectively (R. J. Jaccodine,
Surface energy of germanium and silicon,
J. Electrochem. Soc. **110**, 524 (1963)). The estimates for the III–V compounds are from J. W.
Cahn and R. E. Hanneman, (111) Surface
tensions of III–V compounds and their relationship to spontaneous bending of thin
crystals, *Surface Sci.* **1**, 387 (1964). The numbers for {111} refer to the mean of (111) and
($\bar{1}\bar{1}\bar{1}$) surfaces. The estimates for diamond,
germanium, and silicon are based on cohesive
energies quoted in C. Kittel, *Introduction to
Solid State Physics*, Wiley, New York, 1966,
p. 78.

butions. It should generally be regarded as providing an upper limit
to the value of γ and, indeed, the measured surface tensions (Table
3.1) (p. 56) appear to be consistently less than given by eqn. (5.16).
This model would be expected to give a reasonably good estimate for
materials in which the binding is strongly covalent, and a few calcu-

lated values for crystals with the zinc blende and diamond structure are given in Table 5.2. The values estimated for the (111) cleavage planes of germanium and silicon are in quite good agreement with the experimental values of 1060 and 1230 ergs/cm² respectively.

5.4. METAL SURFACES

The equilibrium structure in the surface region of a metal is of considerable interest in connection with electron emission from metals, electron diffraction, gas–metal reactions, and a number of other phenomena. We shall discuss certain aspects of the structure of metal surfaces and will use some of the concepts that are introduced here in later sections. The structure will also be particularly important in the evaluation of surface energy, surface tension, and other thermodynamic properties.

Qualitatively, in a metal the valence electrons which are responsible for the binding are to a considerable extent delocalized in the crystalline state. For a metal containing small ion cores the wave functions representing the core electrons may be considered essentially unperturbed by condensation from the atomic state, and in this case the bonding energy may be viewed as the sum of a number of terms, viz. the Coulomb interaction among the valence electrons, the Coulomb interaction among the ion cores, the Coulomb interaction between the valence electrons and ion cores, the kinetic energy of the valence electrons, as well as contributions from the vibrations of the ion cores; from these we must subtract the sum of the ionization energies of the free atoms. Even for a uniform crystal (neglecting surface effects) the derivation of the valence electron density distribution which takes all the interactions into account is very complex. This is due to a large extent to the fact that there are correlations in the positions and momenta of the valence electrons. These correlations are partly a result of the Pauli exclusion principle which requires that no two electrons, in the same place at the same time, can be in the same state if we take account of spin. This correlation in the positions of electrons with the same spin leads to a contribution to the potential

energy of an electron called the *exchange interaction* (so called because
the mathematical statement of the exclusion principle is that the total
wave function of the system must be antisymmetric with respect to
the exchange of electronic coordinates including spin). Since electrons repel each other due to the Coulomb interaction, there will also
be correlations in the motion of electrons with different spin. This
type of correlation leads to an additional contribution to the potential energy (relative to the case of uncorrelated motion) known simply
as the *correlation energy*. The exchange and correlation energies are
particularly difficult to treat in a region of rapidly varying electron
density such as occurs at the surface of a metal, and the available theoretical work suggests also that these "many electron effects" are very
important in determining the exact nature of the electron density
variation through the surface region.

A popular model of a metal which has been used extensively in
discussions of the electronic properties of metal surfaces is the so-
called "jellium" model in which the array of positive ions is replaced
by a *uniform* positive charge distribution with charge density equal
to the average for the discrete positive ion lattice. In this model the
surface is then represented by a sharp discontinuity in the positive
charge density with the electronic charge density decreasing smoothly
into the vacuum. The electronic charge density has been calculated
by a number of authors in different approximations, and results have
also been obtained for the 0°K internal energy and electronic work
function of such a surface. It is clear that such a model is not capable
of including surface orientation effects, and some caution must be
used in comparing the results from these calculations with experiments on real surfaces. Nevertheless, many important features of
metal surfaces can be illustrated with this approach. We shall first
review briefly some general features of the free electron model.

5.4.1. *The Free Electron Model*

Consider a cube of monovalent metal of edge dimension L in
which the positive charge is uniformly distributed with density $\varrho_+(r) =$

ne, where *n* is the number of ions per unit volume. If we neglect all many-body effects (the Hartree approximation), the total wave function ψ of the system of electrons can be written as a simple product of one-electron wave functions ψ_i which obey a wave equation

$$-\frac{\hbar^2}{2m}\nabla^2\psi_i + V\psi_i = \varepsilon_i\psi_i, \qquad (5.17)$$

where the ε_i are the one-electron energies and V is the self-consistent potential seen by an electron, i.e. it is the potential due to all the other electrons and the positive charge in the system. If we are interested only in bulk properties, adequate account of the dimensions of the metal are taken by applying periodic boundary conditions, i.e. that the wave functions are assumed to be periodic with period L in the *x*-, *y*-, and *z*-directions which are taken parallel to the cubic edges. It is readily shown that the potential V = a constant, and the plane wave expressions

$$\psi_k = \frac{1}{\sqrt{L^3}}\exp(i\mathbf{k}\cdot\mathbf{r}) \qquad (5.18)$$

for the electron wave functions, where $\mathbf{k}$ is the electron wave vector, lead to a self-consistent solution of 5.17. The electronic charge density corresponding to each plane wave is uniform and thus the electronic charge exactly cancels the positive background everywhere. The constant potential V is indeterminate and is generally set equal to zero.

The possible electron energies (kinetic energies) may be labeled by the wave vectors and are given by

$$\varepsilon_k = \frac{\hbar^2}{2m}\mathbf{k}^2 = \frac{\hbar^2}{2m}(k_x^2+k_y^2+k_z^2),$$

where

$$k_x = \pm\frac{2\pi n_x}{L}, \qquad (5.19)$$

where n_x is a positive integer and similarly for k_y and k_z. The available states are filled (at 0°K) up to a maximum value of k denoted by

9*

k_F (the Fermi momentum) corresponding to a maximum energy ε_F (the Fermi energy) where

$$k_F = (3\pi^2 n)^{1/3}$$

and

$$\varepsilon_F = \frac{\hbar^2}{2m} k_F^2 = \frac{\hbar^2}{2m} (3\pi^2 n)^{2/3}, \qquad (5.20)$$

where n is the mean density of valence electrons; it is often expressed in terms of an effective radius for the valence electrons r_s defined by

$$\tfrac{4}{3}\pi r_s^3 = 1/n, \qquad (5.21)$$

so that

$$k_F = \left(\frac{9\pi}{4}\right)^{1/3} \frac{1}{r_s} \quad \text{and} \quad \varepsilon_F = \frac{\hbar^2}{2m} \left(\frac{9\pi}{4}\right)^{2/3} \frac{1}{r_s^2}. \qquad (5.22)$$

The mean electron density n for most metals lies in the range 1.9×10^{23} to 10^{22} per cm^3 corresponding to r_s values of about 1.0 to 3.0 Å (or about 2–6 atomic units[†]).

5.4.2. The Infinite Square Surface Barrier

The free electron model is not capable of describing the binding of the electrons to the metal. When it is used in connection with surface properties it is therefore necessary to introduce quite empirically a potential barrier at the surface to confine the electronic charge. We will suppose that there is a discontinuity in electron potential described by

$$\left. \begin{array}{ll} V = 0 & \text{(inside the metal),} \\ V = \infty & \text{(in the vacuum),} \end{array} \right\} \qquad (5.23)$$

at the planes $x = 0$ and $x = L$, but will still apply periodic boundary

---

[†] 1 atomic unit = 0.529 Å, the Bohr radius of the ground state of hydrogen ($\hbar^2/me^2$).

conditions at the other surfaces. In this case the wave functions are

$$\psi_k = \left(\frac{2}{L^3}\right)^{1/2} \sin(k_x x) \exp\{i(k_y y + k_z z)\}, \qquad (5.24)$$

where $k_x = (n_x \pi)/L$ with n_x a positive nonzero integer and k_y and k_z take on the same values as for eqn. (5.19). The total electronic charge density is then given by

$$\varrho_- = -|e| \sum_k |\psi_k|^2 = -|e| \sum_{k_x} \left\{\frac{4}{L^3} \sin^2(k_x x)\right\} g_x, \qquad (5.25)$$

where g_x is the number of occupied states with the same k_x. It can be shown that

$$\varrho_- = -n|e|\left(1 + \frac{3\cos X}{X^2} - 3\frac{\sin X}{X^3}\right), \qquad (5.26)$$

where $X = 2k_F x$. The variation of the electronic charge density with

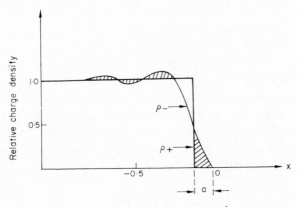

Distance (Fermi wavelengths)

FIG. 5.2. Charge distributions near the surface of jellium with a barrier for electrons of infinite height at the plane $x = 0$. The positive charge is assumed uniform for $x \leqslant a$. Note the oscillation in the electron density into the metal and the existence of a dipole directed into the metal from the vacuum. (After W. J. Swiatecki, *Proc. Phys. Soc. London* A **64**, 227 (1951).) [One Fermi wavelength $\lambda_F = 2\pi/k_F = \pi^{2/3}(\frac{32}{9})^{1/3} r_s$; for $r_s = 4.0$ (the value for sodium), $\lambda_F \approx 7$ Å.

distance for this artificial infinite barrier from eqn. (5.26) is shown in
Fig. 5.2. It increases from zero at the position of the infinite barrier
and exhibits some oscillation about the average bulk value. It should
be emphasized that the solution (5.26) is not a self-consistent one since
we have simply imposed an infinite barrier, and also that exchange
and correlation interactions have been neglected. In the real situation
we should expect the electronic charge density to approach zero
asymptotically as x tends to infinity. This will be the case for any finite
barrier, and it has also been shown that when exchange and correlation
are included the charge oscillations are considerably reduced. It should
be noted that there is an electrostatic dipole associated with the surface and also that in order to conserve charge with an infinite barrier
the barrier would have to be placed at some small distance a (Fig.
5.2) beyond the termination of the positive charge distribution.

5.4.3. Realistic Surface Barrier and the Electronic Work Function of a Metal

The work function Φ of a metal surface may be defined as the differ-
ence between the electrostatic potential energy $-|e|V$ of an electron
just outside the surface and the electrochemical potential of an elec-
tron inside the metal, i.e. $\Phi = (-|e|V - \mu)$. The electron is supposed
removed to a distance small compared to the dimensions of the surface
plane but large compared to atomic dimensions, e.g. 10^{-4} cm. If we
take the electrostatic potential (i.e. the potential seen by a "test"
charge) to be zero just outside the surface[†] (Fig. 5.3), then the work
function for that particular surface is given by

$$\Phi = -\mu. \tag{5.27}$$

There are a number of contributions to the electrochemical potential:

[†] The choice of a zero of electrostatic potential presents some difficulty. With the
convention used here a new reference is required for each different surface of the
metal. If we were to ignore the periodic variation of the potential inside, then the
choice of zero inside the metal would be the most convenient.

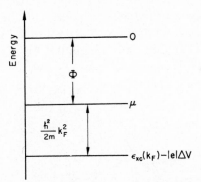

FIG. 5.3. Contributions to the energy of electrons at the Fermi level in a jellium model. The zero of potential corresponds to that of an electron at rest just outside the surface.

(a) the potential energy due to the electrostatic potential ΔV across the surface double layer (i.e. the potential difference seen by a test charge), (b) the exchange and correlation energy $\varepsilon_{xc}(k_F)$ appropriate to an electron with the Fermi momentum in an infinite metal, and (c) the kinetic energy at the Fermi momentum, i.e.

$$\Phi = -\mu = |e|\Delta V - \varepsilon_{xc}(k_F) - \frac{\hbar^2}{2m}k_F^2. \qquad (5.28)$$

The electrostatic potential difference ΔV, the dipole moment P per unit area, and the charge distribution $\varrho(x)$ are related by

$$\Delta V = -4\pi P = -4\pi \int_{-\infty}^{\infty} x\varrho(x)\,dx. \qquad (5.29)$$

In the jellium model with the positive charge abruptly terminated at the plane $x = 0$, the total charge density is given by

$$\varrho(x) = n(e) - \varrho_-(x) \quad \text{for} \quad x \leqslant 0,$$
$$\varrho(x) = -\varrho_-(x) \quad \quad \text{for} \quad x > 0.$$

A number of attempts have been made to derive the electron density in the surface region in the jellium model with the inclusion of exchange and correlation effects, and many of the important features of the results were first demonstrated in the classic paper of Bardeen (see bibliography). A recent self-consistent solution is shown in Fig. 5.4 for different bulk electron densities. At low electron densities the

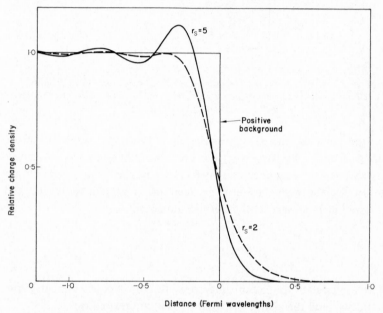

FIG. 5.4. Relative electron density as a function of distance for two different bulk electron densities. Note that at the higher electron densities ($r_s = 2$) the electrons extend further beyond the positive charge and the oscillations in electron density are greatly diminished. (From a self-consistent solution by N. D. Lang and W. Kohn, *Phys. Rev.* B, **1**, 4555 (1970).)

oscillations present in the infinite barrier case still exist, but they are greatly reduced at higher densities. Figure 5.5 shows the corresponding results for the electrostatic barrier ΔV and the total potential energy difference through the surface region. It may be noted that particularly at low densities the normal electrostatic barrier makes only a small contribution to the total potential energy difference, a conclusion also

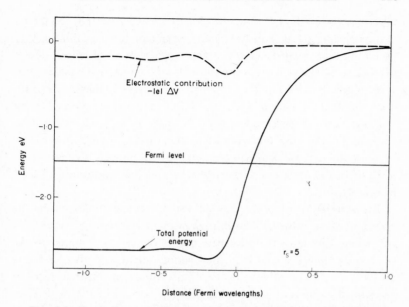

FIG. 5.5. Comparison of the electrostatic contribution to the potential energy of an electron in a free electron metal with the total potential energy including exchange and correlation terms. The total potential approaches the image form $(-e^2/4x)$ outside the metal. (From self-consistent calculations of N. D. Lang and W. Kohn, *Phys. Rev.* B, **1**, 4555 (1970).)

reached in the work of Bardeen. The electrostatic barrier becomes relatively more important at higher electron densities.

The orientation dependence of the work function which is observed experimentally arises from the variation of the electrostatic term, ΔV.[†] To understand this variation we must, of course, go beyond the jellium model and consider explicitly the potential due to the ion cores as well as their arrangement on a lattice. There have been few attempts to take account of the discrete nature of the positive charge in the theoretical evaluation of the work function. What is generally observed

[†] Because of the variation of ΔV with orientation, the electrostatic potential "just outside" different surfaces of the same crystal will be different, and hence electric fields will exist in the vacuum. It is, of course, due to these external fields that the work involved in removing an electron from the Fermi surface and placing it at rest at infinity is independent of the surface through which it is extracted.

experimentally is that Φ is greatest for the most closely packed planes of a particular metal. This trend can be explained in a qualitative way by noting that on surfaces which are rougher on an atomic scale the protrusion of the positive ion cores will tend to decrease the electrostatic dipole barrier which is responsible for the orientation dependence. Smoluchowski has discussed this effect and has shown that the electronic charge distribution near the surface will, indeed, tend to smooth out in directions parallel to the surface mainly due to the decrease in kinetic energy which results. However, a good quantitative analysis of the variation of work function with orientation is not yet available.

The work function of a metal can also, of course, be altered by the presence of an adsorbed layer of molecules either as a result of a permanent dipole of the adsorbed species or due to a change in the surface dipole associated with the bonding interaction. We shall discuss this and some other aspects of adsorption on metal surfaces in section 7.4.

A quantity which is often used in the discussion of potential barriers at crystal surfaces is the *inner potential*, Φ_{inner}. This quantity enters, for example, in the discussion of the diffraction of low energy electrons from surfaces where it plays a role similar to that of the refractive index for electromagnetic waves (see section 6.3). It is the potential energy of an electron inside the crystal relative to one outside. To be more precise the reference point outside the crystal must be specified. It is usually taken to be at a very large distance from the crystal surface so that the inner potential is independent of the orientation of the particular surface being considered. In the jellium model if $\Delta V'$ is the electrostatic potential in the bulk of the metal relative to a point at infinity then the inner potential is given by

$$\Phi_{inner} = -|e| \, \Delta V' + \varepsilon_{xc}(\mathbf{k}). \tag{5.30}$$

The inner potential will depend on the electron energy or more explicitly on momentum due to the momentum dependence of the exchange and correlation contributions to the potential energy. In a real metal the momentum dependence is also due to the interaction with the periodic array of ion cores. Comparing eqns. (5.28) and (5.30) shows that the inner potential and the work function for a surface whose electrostatic dipole barrier is ΔV are related by

$$\Phi_{\text{inner}} = -\Phi - |e|(\Delta V' - \Delta V) + \{\varepsilon_{xc}(\mathbf{k}) - \varepsilon_{xc}(\mathbf{k}_F)\} - \frac{\hbar^2}{2m}k_F^2 \quad (5.31)$$

for a free electron metal.

In treatments of electron diffraction from crystals (see section 6.3) the inner potential is often treated as a complex quantity, the imaginary component arising from possible energy loss processes such as plasmon excitation in the crystal. Thus an electron travelling from the vacuum into a crystal will suffer an increase in its component of momentum normal to the surface due to the real part of the inner potential and a damping of the amplitude of its wave function in the crystal due to the imaginary component. For electrons of energy about 100 eV the attenuation distance is typically of the order of a few angstroms.

5.4.4. Surface Tension of Metals

It is somewhat instructive to evaluate the surface tension at $0°K$ (or surface energy) for the free electron jellium model with an infinite square barrier and with no exchange and correlation interaction. The surface energy is in this case mainly due to the change in kinetic energy of the electrons with a small contribution from the electrostatic energy associated with the surface double layer. A comparison of the results for this model with those for more realistic models helps to indicate the relative importance of the various contributions.

We consider a rectangular block of metal of dimensions $2L$, L, and L in the x-, y-, and z-directions respectively (Fig. 5.6) and imagine the block to be split on the plane $x = 0$. Before splitting the wave functions may be taken to be standing waves

$$\psi_{\mathbf{k}} = \left(\frac{4}{L^3}\right)^{1/2} \sin(k_x x) \sin(k_y y) \sin(k_z z), \quad (5.32)$$

128   PROPERTIES OF CRYSTAL SURFACES

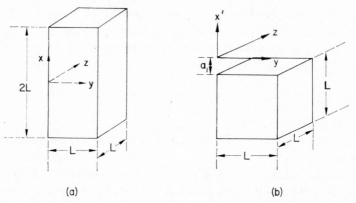

(a)   (b)

FIG. 5.6. After splitting on the plane $x = 0$ the infinite barrier is supposed
to be placed at the plane $x' = 0$, where $x' = (x-a)$.

where

$$k_x = \frac{n_x\pi}{2L}, \qquad k_y = \frac{n_y\pi}{L}, \qquad k_z = \frac{n_z\pi}{L},$$

where n_x, n_y, n_z are positive nonzero integers. The total kinetic energy
E_1 of the electrons in the unsplit block is then given by

$$E_1 = \frac{\hbar^2}{2m}(2)\sum_{k_x}\sum_{k_y}\sum_{k_z}(k_x^2+k_y^2+k_z^2), \qquad (5.33)$$

the summations being taken over all occupied states up to the Fermi
surface. (The factor of 2 arises due to electron spin, there being two
states for each set of momentum components.)

After splitting we imagine the surface at $x = 0$ to have associated
with it an infinite square potential barrier placed at some distance
$x = a$ beyond the positive charge termination (Fig. 5.6). For such a
barrier the wave functions are

$$\psi_k = -\left(\frac{2}{L}\right)^{3/2}\sin(k_x x')\sin(k_y y)\sin(k_z z), \qquad (5.34)$$

where $x' = (x-a)$ and $k_x = (n_x\pi)/L$ and k_y and k_z are the same as in eqn. (5.32). The kinetic energy E_2 of the electrons in each half of the split block is given by eqn. (5.33) for these values of k_x, k_y, and k_z so that the kinetic contribution Δe^s_{KE} to the surface energy is given by

$$\Delta e^s_{KE} = \frac{1}{2} \frac{1}{L^2} (2E_2 - E_1) \tag{5.35}$$

$$= \frac{\hbar^2}{2m} \frac{3n}{2} \left(\frac{\pi k_F}{8} - \frac{4}{15} k_F^2 a \right). \tag{5.36}^\dagger$$

The position of the barrier is determined by the condition that the total electronic charge be equal to the magnitude of the charge of the positive background, i.e.

$$2e \sum_{k} \int_0^L dy \int_0^L dz \int_{-L+a}^0 |\psi_k|^2 \, dx' = L^3 ne. \tag{5.37}$$

This equation determines the parameter a, and this in turn leads to

$$\Delta e^s_{KE} \approx \left(\frac{\hbar^2}{2m} k_F^2 \right) \left(\frac{k_F^2}{16\pi} \right) (1 - 0.78). \tag{5.38}$$

The kinetic energy contribution given by 5.38 is positive for this model and varies as $n^{4/3}$, where n is the bulk electron density; the first term is simply the change in kinetic energy if the infinite barrier is placed at $x = 0$ and is due to the fact that the electrons are confined within smaller volumes. The second term arises from the increase in available volume due to the barrier displacement and for any finite barrier the kinetic energy contribution would be further reduced due to barrier penetration.

The energy stored in the electrostatic dipole contributes to the surface energy per unit area an amount

$$\Delta e^s_{\text{dipole}} = \tfrac{1}{2} \int \phi(x) \{ \varrho_-(x) - \varrho_+(x) \} \, dx, \tag{5.39}$$

† From H. B. Huntington, *Phys. Rev.* **81**, 1035 (1951).

where $\phi(x)$ is the electrostatic potential at x due to the charge distribution. For the charge distribution of Fig. 5.2 the dipolar energy contribution can be shown to be quite small at least for the lower electron densities in comparison with Δe_{KE}^s. For sodium for which $n = 2.5 \times 10^{22}$ per cm³ ($r_s \simeq 4.0$) eqn. (5.38) then predicts a surface energy of about 170 ergs/cm²; this may be compared with a value of about 220 ergs/cm² obtained by extrapolation to 0°K of experimental results on liquid sodium. In view of how unrealistic a model has been used, the agreement is quite remarkable.

Interest in the problem of calculating surface tensions for metals has recently been revived and substantial progress has been made in

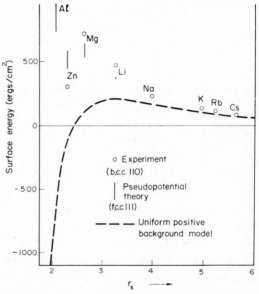

FIG. 5.7. Comparison of the theoretical values of the surface energy with 0°K extrapolations of experimental results (open circles) for liquid metal surface tensions. The dashed curve gives the results for the jellium model. The results obtained from a discrete lattice model using a local "pseudopotential" to represent the ions are indicated by the vertical lines. The upper end of the line is the result assuming a b.c.c. lattice with a (110) surface and the lower end the result for an f.c.c. lattice with a (111) surface. (From N. D. Lang and W. Kohn, *Phys. Rev. B*, **1**, 4555 (1970).)

treating many electron effects. Figure 5.7 shows a comparison of some calculated results for the surface energy as a function of electron density with $0°K$ extrapolations of liquid surface tension data. The dashed curve indicates the surface energy calculated on the basis of a jellium model but with exchange and correlation interactions included in a self-consistent way. The agreement with "experiment" is quite good at low densities but the model fails at higher electron densities where the predicted surface energy becomes negative. It has been shown that this difficulty can be overcome by considering the discrete nature of the positive charge distribution and representing the effect of the ions by a local pseudopotential as indicated in Fig. 5.7.

Some metals such as those of the transition series cannot be described adequately in terms of the free (or nearly free) electron model, and other methods such as atomic orbital techniques will have to be used in treating the surface charge distribution and surface energy. For a discussion of this problem reference should be made to the papers listed in the bibliography.

5.5. ELECTRONIC SURFACE STATES

In discussing the surface energy and surface barriers of metals we have mainly considered the jellium model in which the potential energy of an electron is constant with position in the interior of the metal. In a real crystal the periodic arrangement of the atoms or ion cores leads to a potential for electrons which is three-dimensionally periodic and the allowed energies for electrons moving in such a periodic potential are grouped in bands with forbidden energy zones or band gaps determined by the Fourier coefficients of the potential. If we neglect the finite dimensions of the crystal the wave functions for the electrons, labeled in terms of their momentum $\mathbf{k}$, may be expressed in the form

$$\psi_{\mathbf{k}}(r) = U_{\mathbf{k}}(r) \exp (i \, \mathbf{k} \cdot \mathbf{r}) \tag{5.40}$$

where $U_{\mathbf{k}}(r)$ is a function which has the same periodicities as the lattice and $\mathbf{k}$ is real. Wave functions of the form of eqn. (5.40) but with $\mathbf{k}$ complex are also possible, but these are normally excluded for infinite crystals since they lead to divergent wave functions. However, when the perfect periodicity is disturbed by the creation of a surface or other defect, translational invariance is no longer required and such wave functions can correspond to stationary states of the system. When k is complex it is seen from eqn. (5.40) that the corresponding wave function is an exponentially damped function of position or a so-called evanescent wave. Such wave functions can occur when the lattice is made finite and they are localized in the immediate vicinity of the free surface. The electronic states described by such wave functions are referred to as *surface states*. The calculation of the energies, wave functions, and densities of surface states is an extremely complex problem for any real crystal, and theoretical work on this problem is still in progress. The one-dimensional problem may be discussed with reference to Fig. 5.8 which shows the form of the poten-

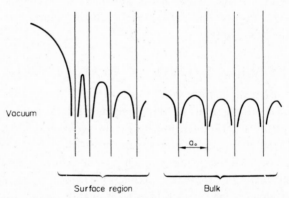

Surface region   Bulk

Fig. 5.8. Possible form of the potential for an electron in a finite crystal in a direction normal to the free surface. The diagram indicates a compression of the lattice planes and a distortion of the potential as the surface is approached. (After D. R. Frankl, *Electrical Properties of Semiconductor Surfaces*, Pergamon, 1967.)

tial of an electron in the vicinity of the surface. The surface states arise from two features—the fact that the crystal is finite and the possible change in the potential near those ion cores which are at or close to the surface plane. Surface states which occur for a pure surface are termed *intrinsic*; states associated with adsorbed impurities are *extrinsic* surface states.

While extrinsic surface states will be important in adsorption on all types of crystals, it is with the semiconductors that both types of surface states can have a very significant influence on the electrical properties associated with the surface. One of the main consequences of the existence of surface states on a semiconductor is the occurrence of a space charge region extending from the surface into the crystal. This is illustrated schematically in Fig. 5.9. Figure 5.9(a) depicts

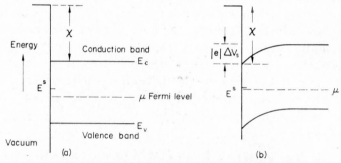

Fig. 5.9. (a) Configuration of the energy bands before introduction of the donor type of surface states at energy E^s for a pure semiconductor. E_v, E_c, μ, and χ are the energies at the top of the valence band, the energy at the bottom of the conduction band (relative to the vacuum level), the Fermi energy, and electron affinity respectively. In (b), which represents the equilibrium situation, electrons have been transferred from the surface states to the conduction band. This leaves the surface positively charged and creates a balancing space charge region across which there is an electrostatic potential difference ΔV_s.

the positions of the valence and conduction band edges and Fermi level before introducing the surface states at energy E^s. It is supposed that these are of donor type, i.e. electrons can be transferred from these states to empty states in the conduction band. At equilibrium some of the surface states will be empty so that the surface is positively charged and there is a compensating negative space charge region across which there is an electrostatic potential difference ΔV_s. For a given distribution of surface states at a particular energy the surface charge and space charge potential difference ΔV_s can be calculated in a fairly straightforward way from the Poisson equation with the use of the Fermi–Dirac distribution. Unfortunately, measurement of these space charge parameters does not allow the surface state distribution to be determined uniquely.

If present in sufficiently large density, surface states can have the important effect of screening the interior of the crystal from its environment. For example, in adsorption, charge transfer involved in bonding can occur between the adsorbing atom and the surface states without directly involving the bulk bands. Charge transfer at semiconductor–metal junctions may essentially involve only the surface states of the semiconductor. It was in fact to explain the behavior of such junctions that surface states were first introduced into semiconductor physics. The literature on the electrical properties of semiconductor surfaces is now very extensive, and the reader is referred to the specialized texts and review articles on the subject for details of the experiments and theoretical situation.

5.6. SURFACE ATOM VIBRATIONS

The presence of a surface influences the modes of vibration of a crystal lattice and modes may exist which correspond to vibrations of the atoms with amplitudes that have the lattice periodicity parallel to the surface but decrease with distance into the crystal. These modes of vibration are in many ways analogous to the electronic surface states mentioned above and are termed *surface phonons*. They can arise in the case of a pure crystal from the change in coordination as well as from the changes in force constants for surface atoms; presumably localized modes will be associated with the various types of surface defects, steps, kinks, vacancies, etc., mentioned in Chapter 4. The adsorption of impurity molecules will also, in general, lead to new modes of vibration which will include the internal vibrations of the adsorbate. The vibrations associated with the surface will be important for a detailed understanding of a number of properties and phenomena. We have already seen that lattice vibrations can make a significant contribution to the surface tensions of crystals. They will be important in several aspects of adsorption—the transfer of the kinetic energy of translation of the incident atom to the lattice, its diffusive motion across the surface, and possible dissociation.

Just as the intensities of beams of diffracted X-rays from crystals are decreased at high temperatures by lattice vibrations (the Debye–Waller effect), so also are the intensities of diffracted beams of low energy electrons which interact primarily with the surface region (see section 6.3). In fact the measurement of the Debye–Waller effect with low energy electrons is one of the most promising ways of studying surface atom vibrations. If we assume that the atoms may be described as harmonic oscillators and consider the diffraction of an electron beam incident normally on a vibrating two-dimensional array, the intensity I of the specularly reflected beam is related to that which would be obtained from the static array I_0 by

$$I = I_0 \exp \left\{ \tfrac{1}{3} \langle u^2 \rangle (\Delta \mathbf{k})^2 \right\}, \tag{5.41}$$

where $\langle u^2 \rangle$ is the mean square amplitude of vibration of a surface atom in the direction of the surface normal and $\Delta \mathbf{k}$ is the scattering wave vector (see section 6.3). The vibration in other directions will be important in determining the intensities of other diffracted beams. A number of studies of the Debye–Waller effect with low energy electrons have in fact been carried out, but because of possible complications from processes of multiple scattering of the electrons the interpretation of the measurements is not yet clear. It is, however, safe to conclude from the observations that the surface mean square amplitudes of vibration along the surface normal are significantly greater (by a factor of about 2) than the bulk values.

The theoretical work on the subject of surface vibrations is of considerable interest. Table 5.3 shows the results of some calculations for the mean square amplitudes of vibrations of atoms in different surfaces of face-centered cubic crystals. These calculations refer to a model crystal twenty layers thick in which the atoms interact through nearest neighbor central forces only. Further calculations including

10*

136   PROPERTIES OF CRYSTAL SURFACES

TABLE 5.3. CALCULATED MEAN SQUARE
AMPLITUDES OF VIBRATION OF ATOMS
AT (100), (110), AND (111) SURFACES
OF FACE-CENTERED CUBIC CRYSTALS

Surface	Normal	Tangential
(100)	2.05	1.55
(110)	2.00	1.64 [1$\bar{1}$0]
		2.15 [001]
(111)	2.05	1.29

The numbers quoted are the ratios
of the mean square displacements rela-
tive to those in the bulk. (From R. F.
Wallis, B. C. Clark, and R. Herman,
Theory of the Debye–Waller factor for
surface atoms, in *The Structure and
Chemistry of Solid Surfaces* (ed. G.
Somorjai), Wiley, 1969, chap. 17.)

next nearest neighbor interactions and noncentral forces apparently
show the same trends. The table indicates that the mean square amplitudes of vibration of surface atoms are considerably greater than those
for atoms in the interior, and also that there is considerably more
anisotropy in the amplitude of vibration of surface atoms.

BIBLIOGRAPHY

BALDOCK, C. R., Determination of the surface energy of a metal by molecular
orbitals, *Proc. Phys. Soc.* A66, 2 (1953).
BARDEEN, J., Theory of the work function: II, The surface double layer, *Phys.
Rev.* 49, 653 (1936).
BENNET, A. J., and DUKE, C. B., Metallic interfaces; II, Influence of the exchange–
correlation and lattice potentials, *Phys. Rev.* 162, 578 (1967).
BENSON, G. C., and YUN, K. S., Surface energy and surface tension of crystalline
solids, in *The Solid–Gas Interface*, Vol. 1 (ed. E. A. Flood), Dekker, New York,
1967, chap. 8.
CYROT LACKMANN, F., On the calculation of surface tension in transition metals,
Surface Sci. 15, 535 (1969).
DAVIDSON, S. G., and LEVINE, J. D., Surface states, *Solid State Phys.* 25, 1 (1970).

DUKE, C. B., Electronic structure of clean-metal interfaces, *J. Vac. Sci. Technol.* **6**, 152 (1969).

EWALD, P. P., and JURETSCHKE, H. J., Atomic theory of surface energy, in *Structure and Properties of Solid Surfaces* (ed. R. Gomer and C. S. Smith), Univ. of Chicago Press, 1953, chap. II.

FRANKL, D. R., *Electrical Properties of Semiconductor Surfaces*, Pergamon, 1967.

HERRING, C., The atomistic theory of metallic surfaces, in *Metal Interfaces*, American Society for Metals, Cleveland, 1952, p. 1.

HUANG, K., and WYLLIE, G. P., On the surface free energy of certain metals, *Proc. Phys. Soc.* A**62**, 180 (1949).

HUNTINGTON, H. B., Calculations of surface energy for a free electron metal, *Phys. Rev.* **81**, 1035 (1951).

LANG, N. D., The density-functional formalism and the electronic structure of metal surfaces, *Reviews of Solid State Physics* (ed F. Seitz and D. Turnbull), Academic Press (1973).

LANG, N. D., and KOHN, W., Theory of metal surfaces; charge density and surface energy, *Phys. Rev.* B, **1**, 4555 (1970).

MANY, A., GOLDSTEIN, Y., and GROVER, N. B., *Semiconductor Surfaces*, North-Holland, Amsterdam, 1965.

MARADUDIN, A. A., Theoretical and experimental aspects of the effects of point defects and disorder on the vibrations of crystals, *Solid State Phys.* **19**, 1 (1966).

MARADUDIN, A. A., MILLS, D. L., and TONG, S. Y., Vibrations of atoms in crystal surfaces, in *The Structure and Chemistry of Solid Surfaces* (ed. G. Somorjai), Wiley, 1969, chap. 16.

SMITH, J. R., Self-consistent many-electron theory of electron work functions and surface potential characteristics for selected metals, *Phys. Rev.* **181**, 522 (1969).

WALLIS, R. F., CLARK, B. C., and HERMAN, R., Theory of the Debye–Waller factor for surface atoms, in *The Structure and Chemistry of Solid Surfaces*, (ed. G. Somorjai), Wiley, 1969, chap. 17.

CHAPTER 6

EXPERIMENTAL METHODS
IN SURFACE STUDIES

6.1. INTRODUCTION

Studies of solid surfaces require rather specialized techniques in addition to those normally used for experiments on the bulk solid state. The reason for this is quite obviously that the methods used for measuring surface properties must either be such that only the topmost layers are sampled or else they should be sufficiently sensitive and accurate that the effect produced by the bulk of the crystal can be subtracted out. Also for fundamental surface studies it is clear that the environment must be carefully controlled since even in a relatively good vacuum of 10^{-7} torr ($\sim 10^{-10}$ atm.) a monolayer of gas could condense on a surface in about 10 sec. Thus most surface experiments (and some industrial processes) require the use of ultra high vacuum (UHV) conditions or very pure inert gas environments. This has been partly responsible for the development of an extensive range of UHV equipment including all-metal valves, ion pumps, sorption, sublimation, and cryogenic pumping devices so that pressures of 10^{-10} torr and lower (until fairly recently attained in only a very few laboratories) can now be achieved with moderate care. We shall not be concerned here with the various recent developments in vacuum technology since extensive accounts can be found in several recent publications.[†] In many instances it is highly desirable to start an experiment with a crystal surface free from adsorbed impurity atoms. The method

† See bibliography.

of achieving this depends on the particular type of crystal involved. The methods that have been used with considerable success include cleavage in UHV, annealing near the melting point in UHV, inert gas ion bombardment, and annealing in a gas which will react with and remove surface impurities. Of these, cleavage is probably the best technique, but unfortunately it can only be used on a relatively few brittle materials and thus is not a generally available technique. It has been used for the preparation of surfaces of a number of ionic crystals, notably the alkali halides, elemental and compound semiconductors, and also a few of the metals including iron, beryllium, and zinc. Annealing at high temperatures is suitable for a number of refractory materials, notably tungsten, but difficulties associated with this technique include the diffusion of impurities from the interior to the surface of the crystal and break-up or faceting of the surface. Bombardment of the surface with a beam of inert gas ions (generally about 10 $\mu A/cm^2$ at 200–1000 V) is quite a generally useful technique and has, for example, been used to produce clean surfaces of aluminum on which there is normally an extremely tightly bound oxide layer. Impurity atoms receive enough energy from the incident gas ion to be ejected (sputtered) from the surface. Some problems may arise from occlusion of the inert gas atoms, preferential sputtering of one of the components in a binary alloy or compound, and change in the surface structure by the bombardment. The choice of a reactive atmosphere to aid in cleaning a surface generally depends on knowing which particular impurity is present. For example, it is frequently possible to remove carbon atoms from a surface by reaction with oxygen in a suitable pressure and temperature range.

In spite of all the difficulties involved, a large number of clean surfaces have now been prepared and studied although in many instances the preparation of the surface constitutes a major fraction of the experimental effort. We shall consider some of the techniques which are particularly useful for surface studies illustrating in most cases the type of information that can be obtained.

6.2. MICROSCOPY OF SURFACES

A wide range of microscopy techniques is now available for studying the topography and structure of crystal surfaces. The available resolution ranges from dimensions of the order of optical wavelengths to angstroms. They include optical microscopy with its numerous variations, transmission electron microscopy of replicas, scanning electron microscopy, electron mirror microscopy, electron emission microscopy, electron field emission microscopy, and field ion microscopy. Which of these techniques may be the most suitable depends, of course, on the type of observation to be made, but it may be remarked that as one seeks higher and higher resolution the limitations on the type of specimen that can be used become more and more severe. Thus whereas optical microscopy can normally be performed on almost any macroscopic specimen, field ion microscopy requires extremely small needle-shaped crystals with radius of curvature at the tip less than about 1000 Å.

6.2.1. *Optical Microscopy*

In the *optical microscope* the lateral resolution is limited by the numerical aperture of the objective lens[†] as shown by the Abbé diffraction theory of image formation. This theory shows that much of the information necessary to produce an image in which very closely spaced objects are resolved is contained in light diffracted at large angles from the specimen. The lateral resolution is then no better than about 0·25 μ and is usually achieved at very small distances between the lens and specimen surfaces. However, by making use of interference effects the vertical resolving power of the optical microscope can be increased considerably, and in one mode of operation vertical displacements of as small as 5 Å can be detected.

[†] Resolution $\simeq (0.61)\{\lambda/(n \sin \theta)\}$, where λ is the wavelength of the incident light, n is the refractive index of the medium between the lens and the object; the quantity $n \sin \theta$ is called the numerical aperture where 2θ is the angle subtended by the lens at the object.

There are three main modes of operation which depend on interference effects—phase contrast, two-beam interference, and multiple beam interference. The method of phase contrast microscopy due to Zernike is used principally in biological studies for examining specimens in which there is very little absorption and hence poor contrast between different parts of the object. The method relies on detecting phase changes produced by different parts of the object, and in the method of Zernike an intensity distribution in the image is obtained which is directly proportional to the phase changes produced by the object. This is achieved by inserting a so-called phase plate into the path of the light passing through the objective to retard or advance the phase of the direct beam relative to that of the scattered light. The technique has been used for examining surface films, but it finds its greatest use in transmission.

The two-beam interference microscopy designed by Linnik for the study of surfaces is an application of the Michelson interferometer. A schematic diagram of the optical arrangement is shown in Fig. 6.1(a). Light from the source S is split by a double right-angled prism and passes through nominally identical objective lenses L_1 and L_2 to be reflected from the specimen surface P and a reference flat surface Q respectively. Provided the reflectivities of the two surfaces are approximately the same and they are effectively in optical contact, superposition of the reflected beams produces a set of interference fringes in which the intensity varies as $\cos^2 \delta/2$, where δ is the phase difference between light from corresponding points of the two objects. Thus the resulting fringe pattern constitutes a contour map of the specimen surface in which the fringe spacing corresponds to a vertical displacement of approximately $\lambda/2$, where λ is the incident light wavelength. It is normally possible to detect fringe displacements on the image of about one-tenth of the fringe spacing, so that with green light illumination the minimum vertical displacement detectable is slightly less than 300 Å. In this method the objective lens can be close to the specimen surface so that high numerical aperture systems can be used with a horizontal resolving power a little less than one micron. Examples of two-beam interference micrographs are shown in Figs.

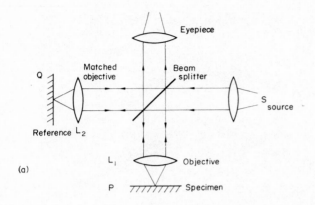

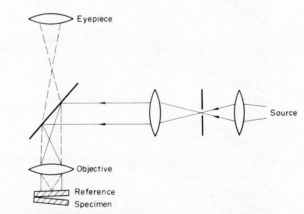

FIG. 6.1. Diagrams showing the basic arrangements for (a) two-beam inter-ference microscopy, and (b) multiple beam interference microscopy. [(a) After S. Tolansky, *An Introduction to Interferometry*, Longmans, 1955. (b) After S. Tolansky, *Multiple Beam Interference Microscopy of Metals*, Aca-demic Press, London, 1970.]

FIG. 6.2. Multiple beam interferogram of a silicon carbide surface showing a spiral system of steps. (Courtesy of S. Tolansky.)

1.12(a) (p. 23)), 7.6 (p. 221), and 7.8 (p. 225). Because of the ease of operation and the relatively good resolving power, the two-beam interference microscope has found wide application even for routine metallographic examination.

The method of multiple beam interference microscopy has been developed principally by Tolansky and is described in some detail in his recent book (see bibliography). It is essentially a development of the Fabry–Perot interferometer in which interference occurs among many beams produced by multiple reflections between a pair of nearly parallel highly reflecting surfaces. In the microscope for the study of surfaces by reflected light it is necessary to place a reflecting reference flat close to the specimen surface as shown schematically in Fig. 6.1(b). When the reflectivities of the two surfaces are very high and there is little absorption, the resulting fringe pattern in reflection is extremely sharp consisting of a set of narrow dark lines on a bright background; an example of such a multiple beam interference micrograph is shown in Fig. 6.2. The great advantage of multiple beam interferometry is its vertical resolution. According to Tolansky, height changes of 5 Å and perhaps less are detectable although horizontal resolution obtainable on a multiple beam interferogram is less than that in the two-beam case. This is due to the fact that multiple beam interference effects cannot be achieved for high angles of incidence such as are involved with the strongly convergent light from a high power objective. The multiple beam method is of particular interest in studying step arrays on single crystal surfaces and finds numerous other applications such as in the examination of surface roughness and measurements of the thickness of thin films.

6.2.2. Transmission Electron Microscopy

With optical methods the lateral resolution is limited to an appreciable fraction of an optical wavelength, and to resolve finer detail than this it is necessary to use radiation of much shorter wavelength. The *transmission electron microscope* employing nearly monoenergetic electrons in the range 50–100 keV (and recently up to 1 MeV) has had its greatest success in examining the internal structure of thin crystalline or biological specimens, and crystal planes of spacing less than 2 Å have now been resolved. The depth of focus in the electron microscope is also considerably greater than in the optical microscope. For example, in an optical microscope operating at maximum resolution the depth of focus is about 0.15 μ; in the electron microscope at a resolution of about 10 Å the depth of focus is about 2 μ. It can thus be used for examining replicas of relatively rough surfaces such as may be produced in fracture. It has also been used for surface studies in a number of ways. The phenomenon of oriented overgrowth of one material on another (epitaxy) has been widely studied with the electron microscope, and experiments have even been performed with-

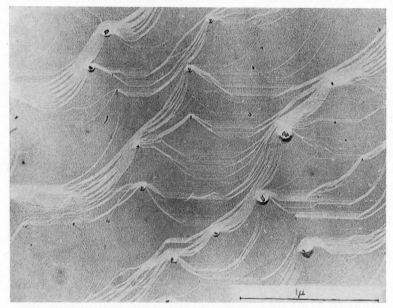

FIG. 6.3. Electron micrograph of a replica of a silicon surface prepared by depositing $\sim$ 1000 Å of silicon on a (111) silicon surface. The replica was prepared by shadowing the silicon surface with platinum at $-150°C$ in a vacuum of $\sim 10^{-7}$ torr and then coating with carbon. Some of the steps are of unit height. (From H. C. Abbink, R. M. Brady, and G. P. McCarthy, *J. Appl. Phys.* **39**, 4673 (1968).)

in the microscope showing directly the nucleation and initial stages of growth of metallic thin films on thin transparent crystals. It has also been widely used for the examination of surface topography by replica techniques which most commonly involve depositing from the vapor on to the surface a thin carbon layer (about 200 Å) which can be removed and examined. Structural details on the surface can be enhanced by shadowing either the original surface or the outer surface of the carbon film with a thin layer (about 10–30 Å) of a heavy metal, commonly palladium or platinum. Maximum resolution on the replica is obtained when the shadowing material is deposited in good vacuum conditions, is of very small grain size, and has low mobility of the surface. An example of a replica of a cleavage surface of silicon prepared at $-150°C$ in a vacuum of $\lesssim 10^{-7}$ torr is shown in Fig. 6.3. In this particular case, monatomic steps are visible on the surface. The best lateral resolution on replicas appears to be about 20 Å, although more commonly 50 Å is about the limit obtainable. Single atomic steps have also been detected in the electron microscope by the method of *decoration* and replication. This is similar in many ways to the conventional preparation of shadowed replicas except that the metal is deposited on the original surface under conditions where the deposition rate is controlled by the rate of nucleation. In general, nuclei tend to form preferentially along steps where the binding energy is presumably higher, and hence the step shows up as a line of particles (Fig. 6.4).

6.2.3. Scanning Electron Microscopy

An instrument which is perhaps more flexible than the ordinary optical reflection microscope or the transmission electron microscope for surface studies is the *scanning electron microscope*. At present lateral resolutions of the order of 100–200 Å can be obtained and depths of field of the order of 1 mm, a feature which allows observation of large changes of height and also allows some surface experiments to be performed while viewing. Further developments may increase the resolution even further. Basically the scanning electron

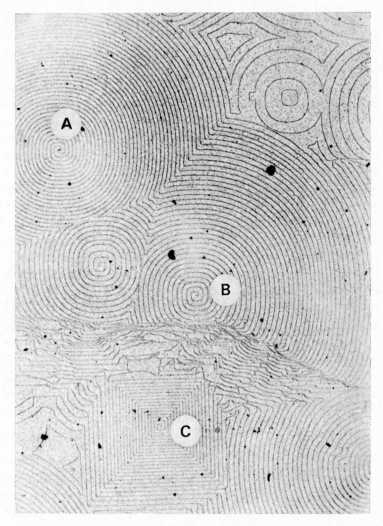

FIG. 6.4. Step pattern on a NaCl (100) surface. The photograph is from a replica of the surface on which the steps have been decorated by gold ncclei.
Some of the steps are of monatomic height 2.81 Å. The step patterns
reveal the presence of screw dislocations intersecting the surface at points
such as *A*, *B*, *C*. (From H. Bethge in *Molecular Processes on Solid Surfaces*,
ed. E. Drauglis *et al.*, McGraw-Hill, 1969.)

microscope employs an electron beam (about 10^{-10} amp at about 20 keV) focused to a diameter of about 100 Å under most favorable circumstances which is scanned across the specimen surface and a collection system to collect the reflected or secondary electrons. This signal is then amplified and used to control the brightness of an oscilloscope beam scanned in synchronism with the electron beam incident on the

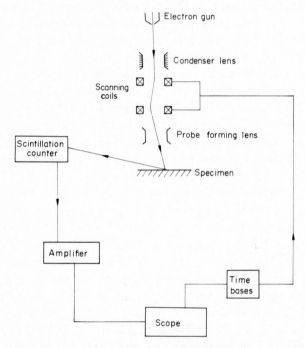

Fig. 6.5. Shematic arrangement of a scanning electron microscope. The primary electron beam is $\sim 10^{-10}$ amp at 20 keV. (After D. Brandon, *Modern Techniques in Metallography*, Buttrwrths, London, 1966.)

specimen. The basic set-up is shown in the schematic diagram, Fig. 6.5. The total back-scattered current received by the collector depends on a number of features of the specimen surface, principally the topography and to some extent the composition and surface voltage. Thus variation of any of these features leads to contrast in the final image. The instrument has been used quite extensively for metallographic examination of fracture surfaces and other metallurgical uses, the detection of defects in microcircuit elements, and also in biological studies. It is now available commercially and is likely to become routinely used for surface examination and possibly even in a production monitor role in the manufacture of microelectronics. It has not yet been used extensively as a quantitative tool in fundamental surface studies, but very important recent developments include the use of UHV chambers and the incorporation of an X-ray spectrometer to analyze the X-rays emitted by the specimen during electron bombardment. By measuring the intensity of particular characteristic X-rays and using this signal to control the intensity on the oscilloscope, it is possible to obtain a distribution of a particular element near the specimen surface. Figure 6.6(a) shows a scanning electron micrograph of a surface of an alloy and Fig. 6.6(b–d) the distribution of some of the elements from characteristic X-ray images. Some measurements have also been made of the intensity of Auger electrons (see section 7.4.1) emitted in the scanning microscope; it is likely that such techniques will be developed to give a spatial distribution of surface elements present in submonolayer amounts. With the incorporation of UHV techniques the scanning microscope should also prove useful in studies of crystal growth and in adsorption experiments; the morphology of adsorbed layers may be visible by virtue of the inhomogeneities in work function which should lead to electrostatic fields very close to the surface.

In the electron mirror microscope a beam of electrons directed at the specimen is reflected at an equipotential surface just outside the real specimen surface; this is achieved by applying to the specimen a potential slightly more negative than the accelerating voltage for the incident beam. The reflected electrons then pass through a system of electron optics[†] to produce a real image on a fluorescent screen. The equipotential surface, at which the electron reflection occurs, is very

[†] A detailed description of electron mirror microscopy and a comparison of different types of electron microscopes are given by A. B. Bok, *A Mirror Electron Microscope*, Delft, 1968.

FIG. 6.6 (a). Scanning electron micrograph of a surface of a pyrophoric alloy (lighter flint!).

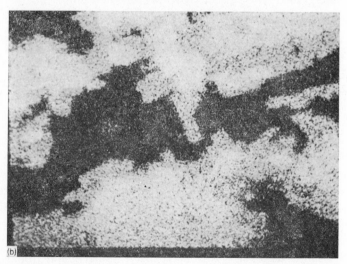

FIG. 6.6(b). This gives an indication of the distribution of some of the elements and is the image obtained from the emitted X-rays characteristic of iron. (Courtesy of Cambridge Scientific Instruments Ltd.)

11*

sensitive to the topography and to the electrostatic potential of the specimen surface. Deviations from flatness, inhomogeneities in potential due perhaps to a nonuniform work function, and also variations in the magnetic field near the surface can all lead to contrast in the

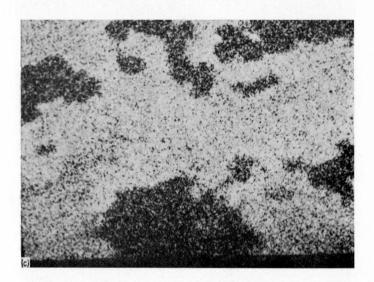

Fig. 6.6. (c). This gives an indication of the distribution of some of the elements and is the image obtained from the emitted X-rays characteristic of cerium. (Courtesy of Cambridge Scientific Instruments Ltd.)

image. This microscope thus offers the possibility of studying a variety of surface features including topography, electrostatic potential distributions (e.g. in microcircuits), and magnetic domain distributions. The resolution of the instrument is of the order of 500–1000 Å.

In electron emission microscopy an image is formed using the electrons emitted from the specimen surface due to heating or to electron or photon irradiation. Contrast in the image results mainly from work function variations over the specimen surface with thermionic and photoemission and also from topographical features in the case of secondary electron emission. The resolution obtainable is of the order of 200 Å.

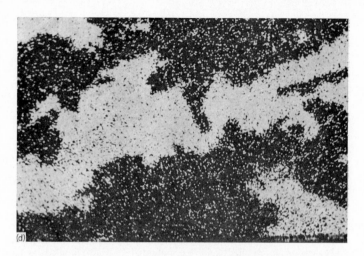

FIG. 6.6. (d). This gives an indication of the distribution of some of the elements and is the image obtained from the emitted X-ray characteristic of lanthanum. (Courtesy of Cambridge Scientific Instruments Ltd.)

6.2.4. Field Emission Microscopy

In section 5.4.3 we have seen that the electrons in a metal are confined by a potential energy barrier which in the absence of any externally applied electric fields is infinite in width. With an extremely large external field the barrier can be made sufficiently narrow that tunneling of electrons into the vacuum becomes appreciable. The electron current density emitted by the metal is a very difficult quantity to calculate exactly. It would require a knowledge of the density of states distribution of the metal, a self-consistent solution of the surface potential variation in the presence of the external field, and then the solution for the tunneling current through this barrier. The solution for the emitted current as a function of the applied field which has been used most extensively is that given by the so-called *Fowler–Nordheim equa-*

tion obtained by assuming a free electron model for the metal, an image type potential for the barrier shape, and evaluating the barrier penetration probability by the WKB (Wentzel–Kramers–Brillouin) method. The potential used is shown in Fig. 6.7; as the applied field increases

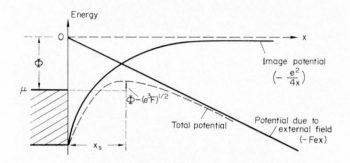

FIG. 6.7. Potential energy–distance curves for an electron near a metal surface. The metal is approximated as being uniform for $x \leqslant 0$. In the absence of an applied field the potential is approximated by the image potential (which will apply at sufficiently large distances). The dashed curve is the net potential in the presence of a strong externally applied field. It has a maximum (or Schottky hump) at $x = \sqrt{(e/4F)}$ so that the barrier height for electrons at the Fermi energy μ is $[\Phi - (e^3F)^{1/2}]$. (After R. Gomer, *Field Emission and Field Ionization*, Harvard University Press, 1961.)

the barrier height and width decrease and since the tunneling probability is such a sensitive function of these quantities the current increases rapidly with applied field. The Fowler–Nordheim equation for the emitted current density in amp/cm^2 takes the form

$$j = 6.2 \times 10^6 (\mu/\Phi)^{1/2} (\mu+\Phi)^{-1} (F/\alpha)^2 \exp\left(-6.8 \times 10^7 \alpha \Phi^{3/2}/F\right), \quad (6.1)$$

where α (of the order of unity) is a relatively slowly varying function of the applied field F (in V/cm), μ the electron Fermi energy, and Φ the work function (both in eV). Substituting typical values of the parameters shows that to obtain an emission current of about 10^2 amp/cm^2

requires a field of the order of about 0.5×10^8 V/cm (0.5 V/Å), an extremely high field indeed. It will also be noted that since Φ and F appear in the exponential the emission will be extremely sensitive to small changes in work function such as may be produced by gas adsorption and to variations in the field at the emitter surface. The necessary fields may be obtained with moderate voltages ($\lesssim 5$ kV) in the field emission microscope (FEM) invented by Müller in 1937. In this the emitter surface is the tip of a very sharp needle-shaped specimen obtained by chemical etching of a thin wire. Because of the small dimensions of the wire the tip is generally a single crystal and is made approximately hemispherical in shape with radius of curvature of about 1000 Å by short time annealing in vacuum. The arrangement of the FEM is shown schematically in Fig. 6.8 and consists essentially of the tip of radius r held at the high positive voltage at the center of a fluorescent screen, radius R. Electrons emitted from the tip thus travel approximately radially and excite the fluorescent screen, producing a so-called field emission pattern of the tip in which the magnification is approximately R/r. The field at the emitter tip is insensitive to the radius of the fluorescent screen and may be found approximately by treating the tip as part of a conducting sphere at voltage V and of radius r for which the radial field at the surface is V/r. In fact the field at the surface of the FEM specimen is somewhat less than this estimate due to the presence of the shank and has its maximum value at the apex.

From eqn. (6.1) it is clear that it is the quantity $(\Phi^{3/2}/F)$ which is important in determining the emission current.[†] Contrast in the image is thus due to its variation over the emitter surface; this arises from local changes in curvature with accompanying distortion of field lines and from the variation of electronic work function Φ with orientation. Planes with high work function will generally appear dark in the image, while sharp protuberances and regions of low work functions will

---

[†] In those metals for which the free electron model is a poor approximation the contrast will be influenced by anisotropies in the density of occupied electronic states.

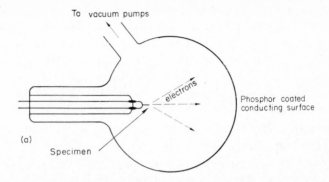

(b)

FIG. 6.8. (a) Diagram of the arrangement of a field electron emission micro-scope. Electrons are emitted from the tip where the radial field is approx-imately V/r and impinge on the fluorescent screen. (After D. Brandon, *Modern Techniques in Metallography*, Butterworths, 1966.) (b) Model of a field emission tip. The model is constructed to approximate as closely as possible to a portion of a sphere but nevertheless shows fairly large sections of certain planes. (Photograph courtesy of A. J. Melmed.)

produce high intensity regions. Figure 6.9 is an example of an electron emission pattern from a nickel specimen with the orientations indicated; it is clear that the emission pattern has the symmetry appropriate to the specimen orientation.

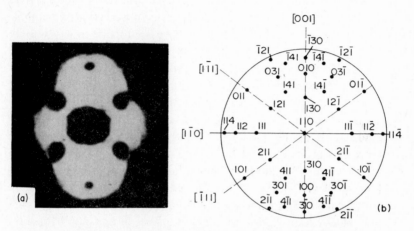

FIG. 6.9. Electron field emission photograph of a nickel single crystal. The orientations are indicated on the corresponding orthographic projection. In general the greatest intensity corresponds to regions of the surface with large step and kink density. The darker regions are low index planes. (From G. Ehrlich in *Metal Surfaces*, American Society for Metals, Cleveland, 1963, p. 230.)

The lateral resolution of the microscope is limited by the magnitude of the component of momentum which the emitted electrons have parallel to the surface. This can arise from the tangential components of momenta possessed by the electron inside the metal, by scattering from variations in the potential parallel to the surface, or it can arise from the spread in momentum as a result of its localization as expressed in the uncertainty principle. It appears that the first effect will in general be most important and that the lateral resolution will be of the order of 20 Å under the most favorable circumstances. Due to the enhanced emission which occurs at the edges of atomically smooth portions of the surface it is sometimes possible to detect steps one interplanar spacing in height normal to the surface (see Fig. 3.5 (p. 62)).

The electron FEM has been used extensively for the study of adsorption, the study of crystal growth from the vapor, the measurement of work functions and their variation with orientation, the study of surface self-diffusion and surface tension, and observation of a number of other phenomena. Some attempts have been made to examine the structure of large molecules adsorbed on the specimen surface, but the results are at present somewhat puzzling; for example, the emission patterns observed do not always have the known symmetry of the molecule. It is probably in investigating the phenomenon of adsorption that the FEM has proved most useful and a large amount of data has now been obtained particularly on the mobility of adsorbed atoms and molecules and the work function changes produced by them. The work function can be obtained from eqn. (6.1) from a plot of $\ln (I/V^2)$ versus $1/V$ (a Fowler–Nordheim plot), where I is the emitted current and V the applied voltage. It is possible to determine the relative work functions of different planes of a crystal by making Fowler–Nordheim plots from measurements of the currents collected from individual planes although the interpretation of such measurements requires some consideration of the variation of the field over the emitter tip. In the case of adsorption on a particular plane, relative work function changes produced by the adsorbed layer can be obtained from Fowler–Nordheim plots before and after the adsorption.

Recently the energy distribution of field-emitted electrons has been studied in considerable detail to obtain information on the density of states in the metal from which they originate and also on the new states (surface states) introduced by adsorption. A discussion of this work may be found in the article by Gadzuk and Plummer listed in the bibliography.

The process of electron field emission is of considerable importance in the initiation of electrical breakdown in vacuum gaps.

6.2.5. *Field Ion Microscopy*

The field ion microscopy (FIM) invented by Müller[†] in 1951 has a resolution of the order of the interatomic distances of close-packed metals. The principles of the microscope are similar in some ways to those of the FEM in that they both involve electron tunneling in the presence of high electric fields, and similar types of specimens are used in the two cases. In the ion microscope, gas atoms are ionized in the immediate vicinity of the specimen surface by field ionization; the ions so created are accelerated and excite a fluorescent screen, which in this mode of operation is at a negative potential relative to the specimen, and produce an image which reflects local variations in the electric field over the specimen surface. The principles of field ionization near a metal surface can be seen with reference to Fig. 6.10. In Fig. 6.10(a) is shown the electron potential energy appropriate to a free atom in the absence and in the presence of an electric field with the ionization level of the atom at energy I below the vacuum level. In the presence of a sufficiently high electric field the potential energy becomes asymmetrical about the core, and there is a finite probability that tunneling will occur from the highest occupied level of the atom to the vacuum, a process known as *field ionization*. The fields necessary for this process to occur must be of the same order of magnitude as those existing in the atom, i.e. of the order of $V/\text{Å}$, which, as we have seen, are obtainable in the immediate vicinity of sharply curved surfaces with voltages of a few kilovolts. Figure 6.10(b) shows the corresponding potential energy diagram for an atom in a strong field near such a metal surface when there is no field penetration into the metal. The net potential energy for the electron is then the sum of the contributions due to the ion core, the applied field, and the interaction with the metal. The diagrams indicate that the potential energy rises from the inner potential of the metal with an image type of variation (see section 5.4.3) and then joins smoothly on to the

[†] The subject of field ion microscopy is reviewed in a recent monograph by E. W. Müller and T. T. Tsong. This book contains extensive references to the literature on FIM and an excellent set of micrographs.

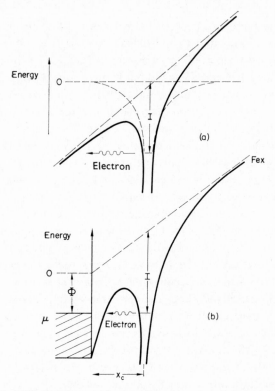

FIG. 6.10. (a) Potential energy for an electron in a free atom without an external field (dashed curve) and in the presence of a large external electrostatic field (solid curve). The dashed line is the potential energy due to the external field alone. In the presence of the applied field the barrier seen by an electron becomes finite in width and is reduced in height. With sufficiently large fields an electron in the first ionization level may escape by tunneling. (b) Potential energy for an electron near a metal surface with an externally applied field F. For the same applied field the tunneling barrier is lower and narrower than in the free atom case due to the image form of the potential near the surface. The diagram shows the gas atom at the critical distance. (Adapted from R. Gomer, *Field Emission and Field Ionization*, Harvard University Press, 1961.)

atomic potential. The transition probability from the ionization level to the metal depends, of course, on the barrier parameters and also on the availability of an empty state at about the same energy in the conduction band of the metal since the transition probability falls off rapidly as the energy difference between the initial and final states increases. For any value of the field there will be a critical distance x_c of separation of the atom and metal below which the field ionization probability falls off rapidly. Referring to Fig. 6.10(b) it is readily seen that $x_c \approx (I-\Phi)e/F$ (the distance where the ionization level of the atom coincides with the Fermi level), neglecting polarization and image effects. For fairly typical conditions of operation of the microscope with helium gas, $I = 24.5$ eV, $\Phi \sim 4.5$ eV, $F \sim 5$ V/Å, the critical distance is about 4 Å. In fact when the field is not too high nearly all of the ionization events are believed to occur within a very narrow zone at distance x_c from the surface since the ionization probability also falls off at greater distances due to the decreased effect of the image type potential in reducing the barrier. Since the applied field is modulated by the lateral local variations in field at the metal surface, and since these will have a repeat distance of the order of interatomic distances, it is clear that x_c must also be of this order of magnitude in order to detect those field variations of the crystal and so resolve atomic detail on the image. From these considerations alone maximum resolution would be obtained at extremely high fields. However, at very high fields two other important effects occur. The ionization probability of a free atom at greater distances from the surface increases so rapidly that few gas atoms get within a distance of x_c before ionization occurs, and at sufficiently high fields the process of field evaporation (see below) occurs. Thus for any particular combination of metal and imaging gas there will be an optimum field.

The resolution of the field ion microscope can also be limited by the magnitude of the component of the velocity of the imaging gas atoms parallel to the surface at the instant when ionization occurs. This effect could be significant since although the normal thermal energy associated with the tangential motion is only of order kT, or about 0.025 eV at room temperature, the kinetic energy, resulting from the attraction of the atom to the tip due to the decrease in polarization energy $(\frac{1}{2}\alpha F^2)$, where α is the atomic polarizability, may be of the order of 0.2 eV. To overcome this effect the kinetic energy of the incoming gas atom must be transferred to the tip so that the effective temperature of the gas atoms becomes that of the tip, a process known as *thermal accommodation*. For this reason the field ion microscope is usually operated with the tip at the temperature of liquid hydrogen (20°K) or liquid helium (4.2°K). It is, in fact, believed that the process of energy transfer occurs in a series of steps in which the gas atom "bounces" off the metal surface, is attracted back to it by the field, and so on, losing part of its energy in each collision. Thus each gas atom may pass through the ionization zone several times as it becomes accommodated to the tip temperature, a feature which substantially increases the ion current.

Figure 4.4 (p. 85) is a field ion microscope picture of an iridium surface clearly displaying the symmetry of the crystal and in which the bright spots correspond to the positions of surface atoms. Not all surface atoms are visible on the micrograph and, in general, those protruding from the surface such as kink atoms or self-adsorbed atoms will give the greatest perturbations in the local field. Figure 4.4 does not represent the equilibium configuration of the surface but has been obtained by the so-called *field evaporation* process. Surfaces which result from heating the tip to temperatures above about half the melting temperature in fact show considerable disorder; the patterns are then rather complex although a quantitative study of them could be profitable from the point of view of obtaining information on the type and concentration of equilibrium surface defects.

The field evaporation process was also discovered by Müller and it is now routinely used in FIM as a means of preparing an essentially perfect crystal surface for use in adsorption, diffusion, and other surface studies. It consists of stripping off atoms from the tip in a controlled manner by application of a large field. The details of the process are not yet fully understood since the exact charge distribution at the surface is not known, but simple models have been proposed which seem to contain the essential principles involved and which can even be used to make quantitative predictions on the necessary fields. The main features can be seen with the help of Fig. 6.11. Here we consider the potential energy of the system as a function of separation of the metal and the ion which is being removed by the field. Curve A (Fig. 6.11(a)) represents a plausible form of the potential

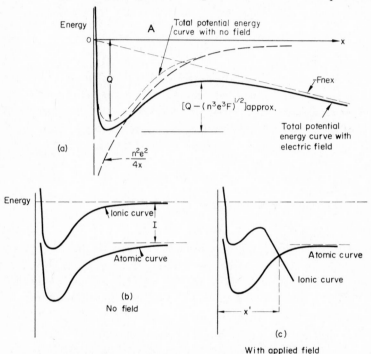

FIG. 6.11. Diagrams illustrating principles involved in field evaporation. (a) The potential energy of an ion of charge $n|e|$ as a function of its distance from a planar metal surface. The attractive part of the potential is approximated by the dashed curve $(-n^2e^2/4x)$ and the binding energy to the surface in the absence of a field is indicated as Q. With an applied field the barrier for escape of the ion from the surface is approximately $[Q-(n^3e^3F)^{1/2}]$. (b) A situation where the stable configuration is that of an atom bound to the surface rather than an ion. In the presence of a strong field (c) the potential energy versus distance curve for the ion is supposed to be lowered below that for the atom beyond some distance x'. Thermal excitation of the atom beyond x' will then allow the possibility of electron transfer to the metal and evaporation as an ion.

energy distance curve in the absence of the field becoming of image type (i.e. $-n^2e^2/4x$) at sufficiently large distance from the surface. In the presence of the field the potential energy exhibits a maximum (the Schottky hump) where

$$\frac{d}{dx}\left(-\frac{n^2e^2}{4x}-neFx\right)=0,$$

i.e.
$$x=\tfrac{1}{2}\,(ne/F)^{1/2}, \tag{6.2}$$

the value at the maximum being $\{Q-(n^3e^3F)^{1/2}\}$. The ion may leave the surface either by receiving enough vibrational energy from the lattice to surmount the barrier or by tunneling through it as in the case of field electron emission. In the former case the mean time τ that the ion will remain on the surface in the presence of the field is approximately given by

$$\tau\approx\frac{1}{v}\exp\left[\{Q-(n^3e^3F)^{1/2}\}/kT\right], \tag{6.3}$$

where v is the frequency of vibrations normal to the surface. For $v\approx5\times10^{12}\ \text{sec}^{-1}$, $\tau\approx1$ sec we require a field of about 3 V/Å for the evaporation at 20°K of a doubly charged ion for which we assume $Q\sim5$ eV. In the case of field evaporation of the crystal itself one may not normally consider the situation as being that of an ion bound to the rest of the metal; this is clearly a limiting case. The other limiting case is that of the removal of an atom, Fig. 6.11(b), (c). Here the potential energy curve for the metal plus neutral atom will lie above that of the negatively charged metal and positive ion at sufficiently large separations as indicated in Fig. 6.11(c). If the atom is thermally excited beyond this distance from the metal, an electronic transition may occur to lower the energy to that of the electronic curve and result in field evaporation. The real situation of field evaporation of a metal surface will be intermediate between these cases and may be viewed as a continuously increasing separation of the ion core from the valence electron background as the field is increased. Fortunately the fields at which evaporation occurs are significantly higher than the optimum imaging field in many systems; there are, however, combinations of crystal and imaging gas where this is not the case, and there is then considerable difficulty in obtaining field ion microscope images. During field evaporation an essentially steady state configuration of the surface or "field evaporation end form" is reached. This is largely due to the fact that atoms in exposed low index planes, for which the value of Q in eqn. (6.3) is high, are much less likely to be removed than those atoms at kink sites around the perimeter of the plane where the binding is smaller and the local field higher.

A large number of experiments on surfaces prepared by field evaporation have now been performed. These include the study of the diffusion of adsorbed atoms and of surface self-diffusion (see Chapter 7), measurements on the binding of adsorbed atoms on different planes, the observation of bulk crystal defects, either thermally produced or induced by particle irradiation, observations on ordered alloys, and more recently the study of the structure of large molecules and structural and compositional changes produced by mechanical contact. The main difficulty in many of these experiments arises from the extremely high field imposed at the crystal surface so that the quantity being studied is influenced by the method of observation. Nevertheless, a great deal of interesting work has been done and methods are being devised to extend the scope of the instrument. Figure 6.12 is an example of the observation of bulk defects by FIM; it shows grain boundary intersections in a W tip.

The high stresses imposed on the tip in the electric field limit the use of the FIM to rather strong solids. For example, in a field of 5 V/Å the stress ($\sim F^2/8\pi$) is about 10^{11} dyne/cm^2, which is of the order of magnitude of the theoretical yield stress (in the absence of dislocations) of most metallic crystals. It appears that little dislocation multiplication occurs in the extremely small tips particularly of the refractory metals, but this can be a problem with softer materials. For these materials different imaging gases and gas mixtures have been used as well as techniques which involve the epitaxial deposition of the solid under study on to a strong substrate tip.

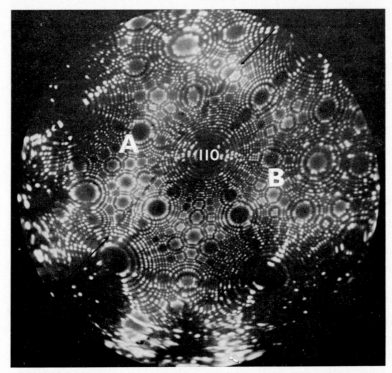

Fig. 6.12. Field ion microscope photograph of a tungsten bicrystal. The position of the grain boundary between crystals *A* and *B* is marked with arrows. (Courtesy of D. M. Davies and B. Ralph.)

Since the field ion microscope resolves atomic detail it is in principle possible to study almost all structural aspects of solids in this way especially when combined with the technique of field evaporation. The most quantitative new information from FIM on surface properties has come from the studies of the binding and migration of adsorbed atoms. The migration studies are discussed in section 7.2. The type of information obtainable on binding is illustrated in Fig. 6.13, which shows results from field evaporation studies on the binding of various atoms to different tungsten surface planes. One of the most exciting recent developments has been the invention of a technique to identify

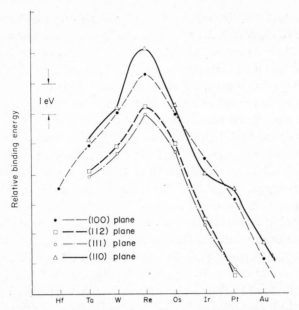

Fɪɢ. 6.13. Relative binding energies of various elements on different planes of a tungsten field ion microscope specimen determined from field evaporation studies. (Courtesy of T. N. Rhodin and E. W. Plummer.)

individual atoms. This "atom probe" analysis, as it is called, consists of identifying individual field-evaporated ions in a mass spectrometer. A small hole in the screen is manoeuvered to coincide with the image of the chosen atom, and the field is raised to evaporate the atom which then passes through the hole to be analyzed in a time of flight mass spectrometer. The technique seems to be particularly appropriate for the study of impurity distributions near lattice imperfections.

6.3. ELECTRON DIFFRACTION
FROM SURFACES

The attenuation distance of a 30 kV X-ray beam incident normally on a crystal surface is of the order of 0.01–0.1 mm. Thus while X-ray diffraction can be used in studies of relatively thick deposits or oxide

12*

layers on surfaces, the technique is not suitable for investigating the structure of clean surfaces or of adsorbed monolayers. On the other hand, 100 eV electrons have attenuation distances of the order of 5–10 Å so that their interaction with crystals is confined to the surface region. The small penetration distance arises from strong elastic scattering by the ion cores and from inelastic electron–electron scattering. According to the de Broglie relation the wavelength λ of electrons with kinetic energy E is approximately $(150/E)^{1/2}$ Å with E expressed in electron volts; the wavelength of a 150 eV electron is thus about 1 Å, and we should therefore expect diffraction in this energy region by the surface layers. Indeed, low energy electron diffraction (LEED) has become an extremely important tool for surface structural studies. It is rather interesting that the discovery of the diffraction of electrons (and hence confirmation of their wave nature) by Davisson and Germer in 1927 was made with electrons of energy 50–500 eV reflected from the surface of a nickel crystal. Following the Davisson–Germer experiment the LEED technique was pioneered mainly by Farnsworth and his co-workers as a method of obtaining surface structures, and in the last 10 years or so has been adopted for the study of a number of surface phenomena. Electrons of higher energy (about 30 keV) have also been used for surface studies when introduced at glancing incidence to the surface plane (reflection high energy electron diffraction) (RHEED). Although the attenuation length is considerably greater at these energies, the penetration into the crystal in the direction of the normal is of the same order of magnitude as in a typical LEED experiment. This can be understood by noting that for an electron of energy 30 keV incident at a glancing angle of about 3° the component of its momentum normal to the surface is about the same as that for a 100 eV electron at normal incidence. Clearly intermediate energies could be used at intermediate angles.

In LEED the relative intensity of particular features of the diffraction pattern depend on the fraction of the surface area occupied by the structure giving rise to those features. On the other hand, because of the glancing angle in RHEED the diffracted beams are extremely sensitive to surface topography. This sensitivity is advantageous in studying such phenomena as the nucleation of oxides, epitaxial growth, and faceting of surfaces, and, in fact, it is possible to obtain information on the morphology of surface nuclei at very small total coverages; however, it also places rather stringent requirements on the initial flatness of the surfaces to be used. We will see later that adsorbed atoms often arrange themselves into ordered structures with two-dimensional unit cells related to those of the substrate. With LEED the periodicities of simple overlayers relative to those of the surface are usually immediately apparent from a single diffraction pattern while with RHEED it is generally necessary to examine at least two diffraction patterns obtained with the beam along different directions in the surface (azimuthal directions). For a complete structural analysis of such overlayers it seems that, in both cases, it is going to be necessary to use the intensities of the various diffracted beams. A satis-

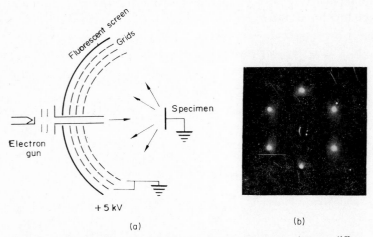

(a)   (b)

FIG. 6.14. (a) Diagram of a LEED apparatus in which the electron diffraction pattern is displayed on a fluorescent screen. Electrons incident on the specimen are diffracted and travel in the field free space to the first grid. The potential of the second grid is held close to the beam voltage so that only elastically scattered electrons (and those that have suffered small losses, e.g. due to phonons) can pass through it. The screen is at a potential of about 5 kV to allow excitation of the phosphor. (b) LEED pattern of a (0001) cleavage surface of beryllium showing the hexagonal symmetry of the basal plane. (Electron energy 86 eV.) ((b) Courtesy of J. M. Baker.)

factory theory for relating these intensities to the crystal parameters has not yet been completely worked out due mainly to the importance of multiple scattering effects. There are, however, noteworthy differences in the two cases. At low energies the electron scattering by the ion cores of the crystal is very strong but only a small number of diffracted beams will be excited simultaneously; at higher energies, although the scattering becomes weaker the shorter wavelength allows a larger number of diffracted waves to be simultaneously excited.

The experimental set-ups commonly used in LEED and RHEED are shown schematically in Figs. 6.14 and 6.15, which also show diffraction patterns obtained by the two methods; Fig. 6.14(b) is a LEED pattern from the (0001) cleavage surface of beryllium displaying the hexagonal symmetry, and Fig. 6.15(b) and (c) are RHEED patterns taken along the [100] and [110] azimuths on a (001) surface of molybdenum. The origin of the diffraction patterns are most easily understood by use of the reciprocal lattice and the Ewald sphere construction.[†] For a three-dimensional crystal defined by the primitive vectors $\mathbf{a}$, $\mathbf{b}$, and $\mathbf{c}$, the corresponding reciprocal lattice is defined by the vectors $\mathbf{a}^*$, $\mathbf{b}^*$, $\mathbf{c}^*$ given by

$$\mathbf{a}^* = 2\pi \frac{\mathbf{b} \times \mathbf{c}}{\mathbf{a} \cdot (\mathbf{b} \times \mathbf{c})}; \quad \mathbf{b}^* = 2\pi \frac{\mathbf{c} \times \mathbf{a}}{\mathbf{a} \cdot (\mathbf{b} \times \mathbf{c})}; \quad \mathbf{c}^* = 2\pi \frac{\mathbf{a} \times \mathbf{b}}{\mathbf{a} \cdot (\mathbf{b} \times \mathbf{c})}. \quad (6.4)$$

For an incident plane wave $e^{i\mathbf{k_0} \cdot \mathbf{r}}$ with wave vector $\mathbf{k_0}$ ($|\mathbf{k_0}| = 2\pi/\lambda$) there will be a set of diffracted waves from the crystal with wave vectors $\mathbf{k}'$ defined by

$$\mathbf{k}' - \mathbf{k_0} = \mathbf{g}, \quad (6.5)$$

where $\mathbf{g}$ is a reciprocal lattice vector (i.e. $\mathbf{g} = h\mathbf{a}^* + k\mathbf{b}^* + l\mathbf{c}^*$, where h, k, and l are integers). In the kinematic or single scattering model, eqn. (6.5) defines the conditions for diffraction maxima and is a statement of the normal Laue diffraction conditions. This can also be shown graphically by the Ewald sphere construction, Fig. 6.16. This

[†] See, for example, B. E. Warren, *X-Ray Diffraction*, Addison-Wesley, 1969.

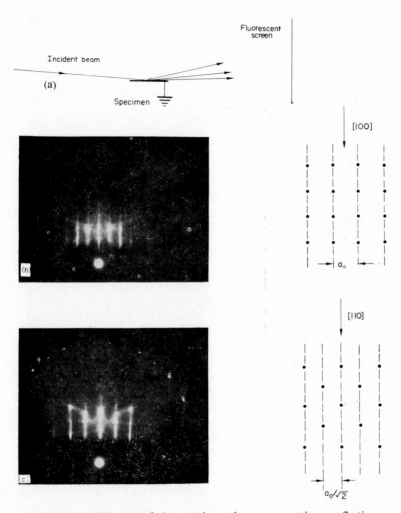

FIG. 6.15. (a) Diagram of the experimental arrangement in a reflection high energy electron diffraction (RHEED) apparatus. The primary energy is generally ~ 30 keV. (b) RHEED pattern of a (001) surface of molybdenum with the incident beam parallel to [100] (referred to as the [10] azimuth). (c) Corresponding pattern for the [11] azimuth. Note the reciprocal relationship between spacing of diffracted beams and interrow spacing on the surface. (Courtesy of H. M. Kennett and A. E. Lee.)

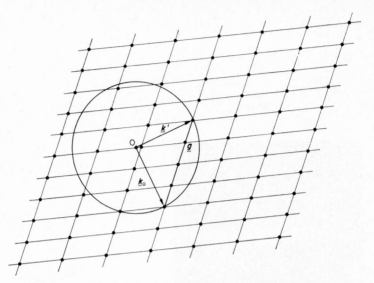

Fig. 6.16. Two-dimensional representation of the Ewald sphere construction. The incident wave vector $\mathbf{k}_0$ is drawn to terminate on some reciprocal lattice point and a sphere is constructed of radius $|\mathbf{k}_0|$ about the origin of $\mathbf{k}_0$. The Laue conditions are then represented graphically by the intersections of this sphere with reciprocal lattice points. For the case shown a diffracted beam will exist in the direction $\mathbf{k}'$.

shows the incident wave-vector $\mathbf{k}_0$ drawn from an origin O to a reciprocal lattice point: with the assumption that the surfaces of constant energy for electrons in $\mathbf{k}$-space are spherical for the crystal being considered, the vectors from the origin to points on the sphere of radius $|\mathbf{k}_0|$ centered on O (the Ewald sphere) represent all wave-vectors for elastically scattered waves. Of these only those which satisfy eqn. (6.5) have nonzero amplitude, so that the possible diffracted beams are represented by wave vectors from the origin to those reciprocal lattice points which lie on the sphere; the diffracted beams in the backward direction are referred to as Bragg maxima and in the forward direction as Laue maxima. In fact the diffraction condition is not quite so strict as suggested by eqn. (6.5) and appreciable intensity can arise in directions corresponding to points on the sphere which lie very close to reciprocal lattice points. The region over which there will be appreciable intensity depends on the number of atoms which scatter coherently; the larger this number the stricter the diffraction conditions. It can be shown, using elementary diffraction theory in which the atoms are treated as independent isotropic scatterers, that the intensity coherently scattered in a direction $\mathbf{k}'$ from a crystal of dimensions $L|\mathbf{a}|$, $M|\mathbf{b}|$, $N|\mathbf{c}|$ is proportional to

$$\frac{\sin^2\left\{\frac{1}{2}L(\mathbf{a}\cdot\Delta\mathbf{k})\right\}}{\sin^2\left\{\frac{1}{2}(\mathbf{a}\cdot\Delta\mathbf{k})\right\}} \cdot \frac{\sin^2\left\{\frac{1}{2}M(\mathbf{b}\cdot\Delta\mathbf{k})\right\}}{\sin^2\left\{\frac{1}{2}(\mathbf{b}\cdot\Delta\mathbf{k})\right\}} \cdot \frac{\sin^2\left\{\frac{1}{2}N(\mathbf{c}\cdot\Delta\mathbf{k})\right\}}{\sin^2\left\{\frac{1}{2}(\mathbf{c}\cdot\Delta\mathbf{k})\right\}} , \quad (6.6)$$

where $\Delta\mathbf{k} = \mathbf{k}' - \mathbf{k}_0$. The conditions for the maxima of intensity are those expressed in eqn. (6.5), and we see also that the intensity falls off about these values of $\Delta\mathbf{k}$ with a rate which depends on the value of L, M, and N. The case $N = 1$ would refer to diffraction from a monolayer crystal defined by the two vectors $\mathbf{a}$ and $\mathbf{b}$. It is clear that the third bracket of the expression (6.6) is then unity for all values of $\mathbf{c}$, and hence the diffraction conditions from such a two-dimensional array of scatters can be represented graphically (Fig. 6.17) as the intersection of the Ewald sphere with the set of lines (or *rods*) in reciprocal space perpendicular to the array each defined by a vector ($h\mathbf{a}^* + k\mathbf{b}^*$). The component of $\Delta\mathbf{k}$ perpendicular to the array, $\Delta\mathbf{k}_\perp$, is quite arbitrary, so that the two-dimensional diffraction condition is

$$\Delta\mathbf{k}_{||} = h\mathbf{a}^* + k\mathbf{b}^* = \mathbf{g}_{||}, \quad (6.7)$$

where $\Delta\mathbf{k}_{||}$ is the component of $\Delta\mathbf{k}$ parallel to the array. For such a two-dimensional lattice the diffracted beams are labeled according to the reciprocal lattice rod intersected by the Ewald sphere as the $(h\,k)$ beam.[†] The diffraction by a crystal surface cannot be represented strictly as a two-dimensional situation since there is appreciable penetration of the incident beam into the crystal. This has the effect of modulating the intensity with varying $\Delta\mathbf{k}_\perp$ as was in fact observed

---

[†] For details of labeling conventions in LEED see the review articles by Lander and by Estrup and McRae and the article by E. A. Wood listed in the bibliography.

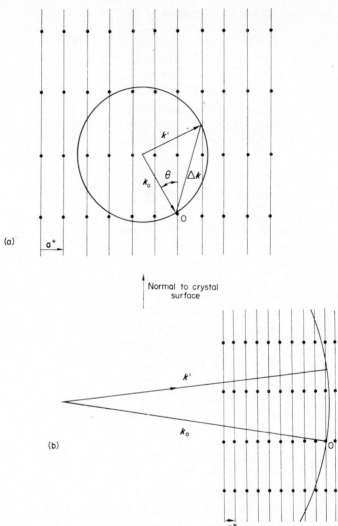

FIG. 6.17. Two-dimensional sections of Ewald sphere constructions appropriate to the LEED and RHEED cases. In (a), the LEED case, the incident beam strikes the surface at near normal incidence. For a strictly two-dimensional lattice, intensity would be observed for all directions such as that of $\mathbf{k}'$, where the sphere intersects the lines drawn through the reciprocal lattice points perpendicular to the surface. The dots correspond to reciprocal lattice points of the real crystal. In (b), the RHEED case, the incident wave vector is nearly parallel to the surface and much greater in magnitude. The lines through the reciprocal lattice rods (perpendicular to the surface) are thus nearly tangential to the Ewald sphere.

by Davisson and Germer in their original electron diffraction experiment; the effect may be observed by measuring the intensity of a particular diffracted beam as the energy of the incident beam is varied and is shown for the case of diffraction from the cleavage surface of zinc in Fig. 6.18. Apart from the complication at about 50 eV, the intensity of the specularly reflected or (00) beam has maxima at about the energies predicted[†] from expression (6.5) for the Bragg peaks from a three-dimensional crystal. In most cases the variation of intensity with energy for a given reflection (i.e. for a particular value of $\Delta \mathbf{k}_{||}$) is considerably more complicated than in the example shown in Fig. 6.18 and generally has maxima in addition to the normal Bragg maxima. While the general features of these curves can be understood on the basis of a multiple scattering or so-called *dynamical theory* which

---

[†] The calculation of the energies of Bragg reflections in the (00) beam involves a knowledge of the inner potential Φ_{inner} (see section 5.4.3). On entering the solid the wave vector of the incident electron is altered due to the inner potential. Thus outside the magnitude of the wave vector $\mathbf{k}_0'$ is given by $(\hbar^2/2m)k_0'^2 = E$, where E is the kinetic energy; inside the wave vector $\mathbf{k}_0$ is given by

$$(\hbar^2/2m)\mathbf{k}_0^2 = (E - \Phi_{inner}) \tag{6.7a}$$

from conservation of energy. The components parallel and normal to the surface are related by

$$\mathbf{k}_{0||} = \mathbf{k}_{0||}'$$
$$\mathbf{k}_{0\perp} = [\mathbf{k}_{0\perp}'^2 - (2m/\hbar^2)\Phi_{inner}]^{1/2}. \tag{6.7b}$$

The effect of the inner potential is to increase the magnitude of the wave vector inside and bring it closer to the normal. The solid therefore has a refractive index for electrons equal to $(E - \Phi_{inner})^{1/2}/E^{1/2}$. The inner potential can be estimated from measurement of the energies at which maxima occur in a particular beam; for the specularly reflected beam the condition for a maximum is from the kinematic theory,

$$2\mathbf{k}_{0\perp} = \mathbf{g}_\perp$$

o

$$E \cos^2 \theta - \Phi_{inner} = (\hbar^2/2m)n^2(\pi/d)^2, \tag{6.7c}$$

where n is the order of the reflection from the the set of planes of spacing d which are parallel to the surface and θ is the angle which the incident wave vector makes with the surface normal. A plot of $E \cos^2 \theta$ versus n^2 should be a straight line whose slope gives the interplanar spacing d and with intercept Φ_{inner}. Inner potentials for a number of materials have been determined in this way; generally Φ_{inner} varies with incident energy.

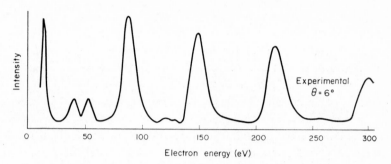

FIG. 6.18. Measured intensity of the specularly reflected beam as a function of electron energy from a (0001) cleavage surface of zinc. The angle of incidence is about 6° from the normal. The main peaks may be identified as Bragg reflections from the three-dimensional crystal. The width of the peaks appears to be determined by the attenuation caused by inelastic loss processes. The inner potential can be determined from the observed positions of the Bragg peaks as indicated in footnote †, p. 175.

includes inelastic scattering of electrons, the analyses have only recently reached the point where intensity data can be used in surface structure determinations. Nevertheless, the symmetry and dimensions of the unit cell of two-dimensional structures on surfaces can often be determined from observations of the diffraction pattern alone. For this reason the technique of LEED has been used extensively for the study of adsorbed monolayers and it has been shown that in most cases of chemisorption on to clean single crystal surfaces the adsorbed layers exhibit long range periodicities related to those of the substrate. Figure 6.19 shows an example of a (110) nickel surface with the formation of a superlattice accompanying oxygen adsorption. The superlattice is labeled according to the dimensions of its two-dimensional unit cell or unit mesh relative to that of the substrate surface plane. In Fig. 6.19 additional diffracted beams occur at the reciprocal lattice positions $(h\pm\frac{1}{2}, k)$ so that one of the primitive vectors of the superlattice is twice as large as that of the substrate; the overlayer is referred to as a "primitive two by one" and written $p(2\times1)$. It is clear that with this particular structure there are two possible equivalent positions of the overlayer relative to the substrate so that ordered

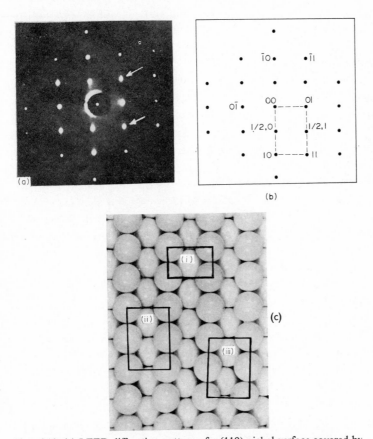

FIG. 6.19. (a) LEED diffraction pattern of a (110) nickel surface covered by
$\frac{1}{2}$ monolayer of oxygen atoms taken at near normal incidence with electrons
of energy 135 eV. The reflections of the type marked with arrows are due to
the overlayer; the other spots are also characteristic of the clean surface.
(b) indicates the way in which the reflections are labeled, i.e. in terms of
multiples of the reciprocal space vectors describing the substrate unit mesh.
(c) A ball model of a clean (110) surface of a face-centered cubic crystal.
The surface mesh (i) is indicated as well as two possible equivalent positions (ii) and (iii) of the unit mesh of the ordered adsorbed layer. (Photographs courtesy of L. H. Germer.)

domains with antiphase boundaries are likely to occur. These seem to have been observed in a number of systems. Similar types of super-lattices have also been observed by RHEED although it is only rela-tively recently that UHV conditions have been used in this type of apparatus.

Low energy electron diffraction has now been used to study a num-ber of surface phenomena including epitaxial growth, faceting, step distributions, thermal expansion, and surface atom vibrations. The last of these applications consists of measuring the decrease in intensity of diffracted beams with increasing temperature due to scattering of electrons by phonons—the Debye–Waller effect. The preliminary results from these experiments indicate that the amplitude of vibra-tion of surface atoms along the surface normal is significantly greater than that of bulk atoms although, as indicated earlier, it is not yet clear what effect multiple scattering has on the observed temperature variation of intensity so that quantitative results are not yet available. Static displacements of the last few layers, of the type discussed in sections 4.3.1 and 5.2, have also been sought by diffraction tech-niques. However, it is still difficult to separate effects associated with variations in the inner potential with energy, details of the surface barrier shape and with multiple scattering from those associated with changes in interplanar spacing as can be seen by examining eqn. (6.7c). At present a number of researchers in the field are developing experimental systems to observe with improved energy resolution the diffracted or elastically scattered electrons in the very low energy region ($\lesssim 10$ eV). In this energy region effects due to the surface po-tential barrier should become important, and it may be possible to derive information on the surface charge distribution from such studies.

6.4. SURFACE COMPOSITION

For a basic understanding of almost any surface phenomenon it is necessary to know the chemical composition of the last few layers; the distribution of components with depth into the crystal would also be of considerable interest in many cases. This type of analysis is, of course, difficult because of the rather small number of atoms which are involved and because of the problems associated with separating out effects which are due to the underlying crystal. Until relatively recently, techniques for making such analyses were not available and, indeed, the influence of trace amounts of impurities is unknown in many experiments. The problem of surface chemical analysis is now receiving a great deal of attention, and a number of very elegant and potentially accurate methods are being explored. It is rather difficult to predict which of these will prove to be the most generally useful, and the choice will often be dictated by such considerations as the specimen geometry and experimental environment. For example, the most elegant method is probably that of "atom probe analysis" in the field ion microscope (section 6.2.5) in which individual atoms can be identified, but this is, of course, restricted to the needle-shaped specimens required for that instrument. The majority of surface experiments are performed on more macroscopic, usually flat, specimens, and it is for these that most of the techniques have been developed. We consider here some basic aspects of the various methods. They can be grouped according to whether identification is based on the electronic energy levels or on the mass of the atoms or ions. Within the first group are the techniques of (i) *photoelectron and secondary electron spectroscopy*, (ii) *X-ray appearance potential measurements*, (iii) *electron probe microanalysis*, and (iv) *Auger electron spectroscopy*. The second group includes *mass spectrometer identification* of atoms or ions removed from the surface by (i) *thermal desorption*, (ii) *ion bombardment*, or (iii) *field desorption* (atom probe analysis), and also a technique based on the measurement of the *energy losses of inert gas ions backscattered from the surface*. In addition to these two main groups of techniques it may also be possible in special circumstances to obtain some information on surface impurities from the use of radioactive tracers, electron diffraction analysis, or work function measurements.

6.4.1. *Methods Based on Electronic Energy Levels*

The various methods within this group may be discussed with reference to Fig. 6.20 which depicts the energy levels of an atom. Since each atom has a unique set of available energy levels it may be identified by the ejected electrons or photons due to transitions induced by an incident X-ray or electron.

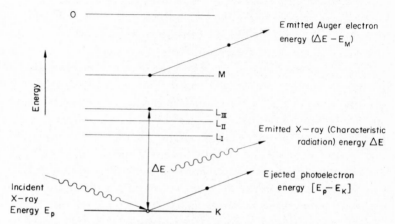

FIG. 6.20. Schematic energy level arrangement in an atom with some of the possible transitions that can occur. An incident photon of energy E_p may eject an electron from an occupied state, e.g. the state labeled K with kinetic energy $[E_p - E_K]$ causing ionization. This initial excitation may also be produced by an incident electron. De-excitation can occur in two different ways. An electronic transition can occur from an upper level with the emission of an X-ray photon. Alternatively, the energy ΔE released in the transition can be coupled to another electron such as that in the M level which is ejected with energy $(\Delta E - E_M)$. The latter mode is known as an Auger process, and would be labeled in this case as $KL_{\mathrm{III}}M$. Following the initial ionization event, all of the energy levels will in fact be displaced to some extent.

Consider the case of X-ray excitation. If the primary X-ray is essentially monochromatic, of energy E_p, the distribution in energy of the photoelectrons produced by direct ionization will have peaks at the energies $(E_p - E_i)$ where E_i are the binding energies in the various atomic levels. An energy analysis of the emitted electrons thus serves to identify the type of atom. This is the basis of the method known as ESCA (electron spectroscopy for chemical analysis) developed by Siegbahn and co-workers and has been used for the study of the energy levels of free atoms and molecules as well as solids. The technique requires rather sophisticated instrumentation (shown schematically in Fig. 6.21(a)) to obtain adequate energy resolution; it uses X-ray characteristic radiation as the primary exciting beam and a high resolution electron spectrometer for analysis of the emitted electrons. A notable feature of the type of spectrum obtained is the sharpness of the emission peaks. This is due primarily to the fact that the major energy loss process experienced by the escaping photoelectrons is that of plasmon excitation which requires energy increments of $\gtrsim 10$ eV. For maximum ionization efficiency the primary energy should be about three times the ionization energy, the ejected photoelectrons under these circumstances having energies of about $2E_i$. For an atom such as silicon the electron ejected from the K-shell would thus have

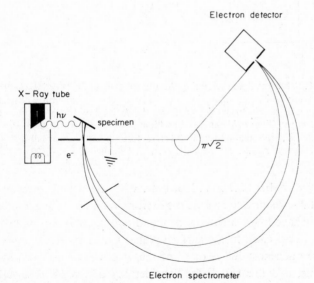

FIG. 6.21(a). Schematic view of an arrangement used in ESCA for the analysis of electrons ejected by X-rays.

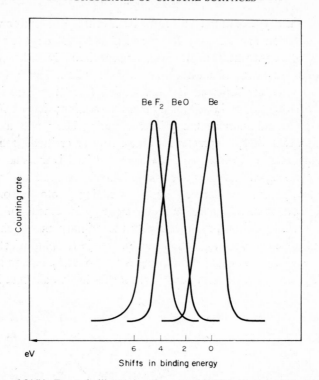

FIG. 6.21(b). Example illustrating the use of ESCA for detecting electron energy level shifts accompanying chemical compound formation. The peaks correspond to electrons ejected from the *ls* level of Be in the metallic form, as beryllium oxide and as beryllium fluoride. (From K. Siegbahn *et al.*, Atomic, molecular and solid state structure studied by means of electron spectroscopy, *Nova Acta Regiae Soc. Sci. Upsaliensis*, ser. 4, **20** (1967).)

an energy of about 3.7 keV. However, with the use of lower energy incident X-rays or electrons as the exciting radiation, the ejected electrons will have lower kinetic energy and will be detected only if they originate very close to the surface plane. Under most circumstances only the outermost 100 Å or so of the solid will be sampled by this technique, and this should be considerably decreased by a suitable choice of incident energy. Apart from being an analytical tool this type of electron spectroscopy has been used to study shifts and broadening in electronic levels accompanying the formation of chemical compounds and condensation to the solid state (Fig. 6.21(b)). Similar studies promise to be extremely valuable for the case of atoms chemisorbed on surfaces.

The other methods involve the detection of the radiation accompanying de-excitation of the atom. Having created a vacancy in a low-lying level by the incident beam, a transition can occur to this state from a higher energy level with the release of an amount of energy ΔE (Fig. 6.20); the transition can be accompanied by the emission of an X-ray of energy ΔE (characteristic X-ray) or an electron can be emitted from another state, e.g. the M-level (Fig. 6.20), with kinetic energy approximately[†] equal to $(\Delta E - E_M)$. The latter is called an Auger process and the emitted electron an Auger electron.

In the X-ray appearance potential measurements the exciting radiation is an electron beam. As the energy of this beam is increased through that of one of the electronic levels, ionization of that level becomes possible and a new set of transitions can occur. Thus at the various ionization energies there should be stepwise increases in the X-ray output detectable with a suitable photomultiplier. In practice the sensitivity is increased considerably by modulating the energy of the incident electrons and detecting synchronous modulations in the X-ray output. The sensitivity to the surface region in this method arises from the use of low energy primary electrons ($\lesssim 1$ keV) which can be introduced at glancing incidence. Although this method is still in the fairly early stages of development it is particularly attractive in that it involves relatively simple instrumentation.

In electron probe microanalysis the energy of the incident electron beam is held constant and the intensities of the emitted X-rays are measured with suitable X-ray spectrometers. Normally the incident beam has an energy of about 10–30 keV and is focused to a diameter at the specimen of about 1 μ. The depth of specimen which contributes to the emitted X-rays is also of the order of a micron and even

---

[†] In calculating the energies of emitted Auger electrons or characteristic X-rays it is in fact necessary to take account of the shift in levels following the initial ionization.

when the beam is focused to extremely small diameters (about 200 Å) as in the scanning electron microscope it is still in excess of about 0.1 μ. The depth sampled is essentially fixed by the primary energy which determines the depth at which X-rays are generated. The lateral resolution of about 1 μ in composition variations over the surface makes the technique extremely useful for studying composition profiles across semiconductor junctions, precipitates, etc.

The method of Auger electron spectroscopy is at the moment the one which has been used most extensively for the study of the composition of single crystal surfaces. This is largely due to the relatively simple instrumentation required to give adequate sensitivity and to the fact that there are generally Auger electrons with energies below about 500 eV. The depth of specimen which is sampled by the technique is determined by the escape depth of the Auger electrons and is believed to be about 3–5 atomic layers[†] with a sensitivity of about 0.01 monolayer under the most favorable conditions. The energy spread of the Auger electrons does not depend on that of the primary beam but is determined by the width of the electronic energy levels (or energy bands in the case of solids) which are involved and the energy losses during escape. Thus no special precautions have to be taken with regard to the source, and a simple electron gun with stable emission will suffice. The technique is rather conveniently combined with the type of LEED apparatus shown in Fig. 6.14, where the grid system of the LEED optics is used as a retarding potential analyzer although other types of analyzer with higher resolution have also been used. Figure 6.22 shows an Auger spectrum from a nickel surface showing in addition to nickel peaks others which are characteristic of sulfur and carbon, present as impurities. A considerable effort is presently being made in calibrating spectra to obtain absolute concentrations of surface elements. It may be noted that for the low atomic number elements ($\lesssim 30$) the process of Auger electron emission is the dominant method of de-excitation following ionization. This is one of the main reasons that the light elements are difficult to detect by X-ray fluorescence.

[†] The mean free path for inelastic scattering of electrons is energy dependent. For the common metals it has a minimum in the region of 100 eV.

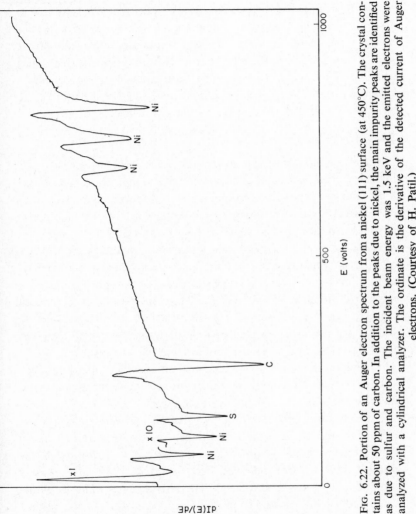

FIG. 6.22. Portion of an Auger electron spectrum from a nickel (111) surface (at 450°C). The crystal contains about 50 ppm of carbon. In addition to the peaks due to nickel, the main impurity peaks are identified as due to sulfur and carbon. The incident beam energy was 1.5 keV and the emitted electrons were analyzed with a cylindrical analyzer. The ordinate is the derivative of the detected current of Auger electrons. (Courtesy of H. Patil.)

It will be clear that various combinations of these techniques may be used to advantage. For example, a system for analysis of Auger electrons generated in the scanning electron microscope has been used. This is potentially capable of combining the high lateral resolution of the scanning microscope with a depth resolution of a few atom layers to give a detailed spatial distribution of surface impurities. No doubt other combinations will be useful for surface analysis, and we should anticipate results on surface binding from studies of the shift of the electronic levels of surface atoms from those of the free atoms.

6.4.2. Methods Based on Identification by Mass

The techniques within this group differ from those mentioned above in the important respect that the method of analysis generally produces a net change in the surface composition. Those techniques which use mass spectrometers necessarily require removal of atoms or molecules from the surface. In the thermal desorption method the specimen temperature is increased sufficiently that the probability of a surface atom or molecule receiving enough kinetic energy from lattice vibrations to overcome its binding to the crystal becomes appreciable. The method is suitable only for volatile components and has been used mainly in connection with the study of adsorbed gases (see section 7.4). It is generally performed by rapidly raising the crystal temperature (flash desorption) so that appreciable pressure increases occur. The method is useful in checking for the presence of a particular volatile species; it is relatively straightforward, but cannot really be regarded as a general analytical technique.

Recently a new analytical tool known as the *ion microprobe* has been developed and promises to be extremely useful for surface studies. This basically involves bombarding the surface with a beam of gas ions (of about 10 keV) and detecting the sputtered material after mass separation in a mass spectrometer. The instrument is rather sophisticated and can be modified to produce an "ion image" of the surface (or a lateral distribution of a component) with a resolution of the order of $1\,\mu$ as well as a distribution of a component with depth. In one form a spatial distribution of an element is obtained by ras-

tering a focused beam of diameter of the order of 1 μ across the surface in a manner similar to that of the scanning electron microscope.
The ions sputtered from the surface are separated in a mass spectrometer and the current of a particular type of ion is used to control
the brightness of an oscilloscope beam rastered in synchronism with
the ion beam. Because of the extremely high sensitivity of the mass
spectrometer, concentrations in the part per billion range are detectable. The method is extremely sensitive to the surface layers and has
the potential of detecting impurities present in concentrations which
constitute a very small fraction of a monolayer as well as providing a
distribution with depth, with a resolution of the order of a few interplanar spacings, as the crystal is eroded away by the primary beam.
At present it is not generally possible to give quantitative analyses
since, for a particular bombarding gas, the sputtering efficiency varies
widely from one element to another and also depends strongly on the
chemical state of an element.

The experimental arrangement for the low energy ion scattering
method of detecting surface components is shown schematically in
Fig. 6.23(a). A primary beam of nearly monoenergetic ions (usually
about 1 μA/cm^2 at about 0.5–2 keV of inert gas ions) is incident on
the specimen surface. Some of these are scattered by the surface without neutralization and the scattered current, in a particular direction,
is measured as a function of energy with a suitable energy analyzer and
detector. The interesting feature of the scattered ion current is the
occurrence of peaks corresponding to energy losses of the primary
beam which are characteristic of the masses of the surface atoms.
Furthermore, it appears that the values of the energies of the peaks
are described rather well by treating the collision as a simple two-
body elastic interaction between the incident ion and the surface atom
assuming the latter to be completely decoupled from the rest of the
crystal during the collision. Thus a surface atom of mass m_2 produces
a scattered ion current peak at an energy E given by

$$E = E_0 \left(\frac{m_2 - m_1}{m_2 + m_1} \right) \quad \text{provided } m_2 > m_1 \tag{6.8}$$

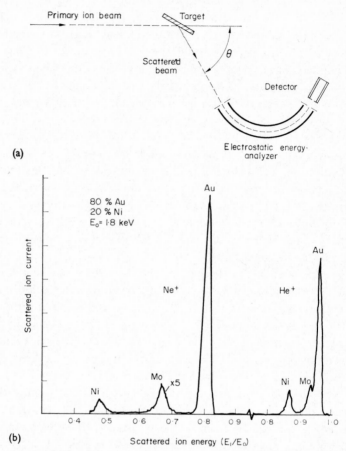

(a)

(b)

FIG. 6.23. (a) Basic arrangement for the study of surface composition by low energy ion scattering. (b) Distribution of scattered ions from the surface of a Au–Ni alloy. When $\theta = 90°$ the peaks occur at energies $E_0\dfrac{(m_2-m_1)}{(m_2+m_1)}$, where m_2 is the mass of the atom at the surface, m_1 that of the incident ion and E_0 the primary energy. The results are given for the scattering of both neon and helium ions. The distributions indicate the presence of molybdenum as an impurity on the surface. (From D. P. Smith, *Surface Sci.* **25,** 171, 1971.)

for a scattering angle of $90°$, where E_0 is the primary beam energy and m_1 is the mass of the primary ion. An example of the type of distribution of scattered ions obtained is shown in Fig. 6.23(b) for the surface of a Au–Ni alloy. It appears that only the outermost one or possibly two layers determine the scattered current distribution. As with all methods for surface analysis, there is the difficulty of obtaining suitable standards against which the technique can be calibrated, and it should be regarded as being semiquantitative at present. Like the ion microprobe there is the possibility of analyzing the depth distribution of a component since the primary ion beam serves to sputter the surface at a controlled rate.

6.5. WORK FUNCTION MEASUREMENTS

In section 6.4.3 we have seen that the work function Φ for a particular crystallographic plane is the difference between the Fermi energy for electrons in the crystal and the electrostatic potential energy just outside that plane and we have discussed the various factors which contribute to the work function and the surface barrier in the case of a metal. We shall now outline briefly some of the methods for measuring Φ.

The work function of metal surfaces is the parameter which is or prime importance in electron emission phenomena from solids and is thus of practical importance in assessing the usefulness of a particular material as a thermionic emitter or photosensitive detector. Although the work functions for metallic emitters have values which only range over about a factor of two, the corresponding spread in emission currents at a particular temperature is large because of the exponential dependence of the emission current on Φ. The work function is also of considerable interest in fundamental surface studies because of its great sensitivity to the state of the surface, particularly the presence of adsorbed gases or segregated bulk impurities. The modification of the surface charge distribution by the adsorption of gas atoms or molecules leads to a net change in the work function which is often referred to (somewhat misleadingly) as the *surface potential* of the adsorbing species at a particular coverage. The work function change may result from the permanent dipole moments associated with the adsorbed molecules as well as from the moment associated with the bonding.

There are two main types of work function measurements.[†] One depends on detecting the electron current emitted from the surface as the result of an increase in temperature, incident radiation, or the application of high electric fields. These methods give absolute values of the work function. The second group involves the measurement of the potential difference existing between the surfaces of two materials which are electrically connected—*the contact potential difference*—and therefore give only relative work function values. This group includes the Kelvin method, the retarding potential method, and the saturated and space charge limited diode methods.

6.5.1. *Absolute Measurements*

(a) *Thermionic emission*

The electrons in a metal obey the Pauli exclusion principle, and the occupancy of the available states of energy ε is described by the Fermi–Dirac distribution

$$f(\varepsilon) = \frac{1}{\exp\{(\varepsilon - \mu)/kT\} + 1},\qquad(6.9)$$

where μ is the electrochemical potential or Fermi level of the electrons. $f(\varepsilon)$ is essentially unity up to energy values close to μ and then falls to zero with an exponential or Boltzmann tail. In fact for values of ε which lie above the vacuum level, i.e. $\varepsilon = (\mu + \Phi) + \varepsilon'$, eqn. (6.9) becomes

$$f(\varepsilon) \approx \exp(-\Phi/kT)\exp(-\varepsilon'/kT).\qquad(6.10)$$

Since the work function Φ is ordinarily of the order of a few electron

---

[†] A comprehensive listing of techniques and results is given in a recent article by J. C. Rivière. This article contains a fairly detailed description of the various experimental arrangements.

volts, the occupation probability is extremely small for positive values of ε' (e.g. for $\Phi = 4$ eV, $T = 10^3°$K, $\varepsilon' = 0$, $f(\varepsilon) \sim 10^{-21}$). Nevertheless, the density of available electronic states at the vacuum level may be sufficiently high (e.g. about $10^{22}/cm^3/eV$ in a free electron metal) that the total number of electrons with energy above the vacuum level is appreciable so that significant currents could be thermally emitted when a small external field is applied. In calculating the current density emitted by a particular surface (taken to be the number of electrons with an outward component of momentum and with $\varepsilon' \geqslant 0$) it is necessary to use the relationship between the electron energy and its momentum. For the case of a free electron metal (see section 5.4.1) the problem can be solved exactly to give

$$j = AT^2 \exp(-\Phi/kT) \qquad (6.11)$$

where $A (= 4\pi m k^2 e h^{-3})$ has the value 120 A/cm^2/deg^2; eqn. (6.11) is known as the Richardson–Dushman equation. It neglects electron reflection at the boundary, possible temperature variation of Φ, and assumes that there are no external fields which may change the barrier height. The determination of the work function then consists of measuring the emitted current (extrapolated to zero external field if necessary) as a function of temperature; Φ is usually obtained from a plot of $\ln(j/T^2)$ versus $1/T$ (Fig. 6.24), a Richardson–Dushman plot.

It may be noted that for a nonuniform surface, e.g. that of a polycrystalline ribbon, in which the work function varies over the surface area, each patch of the surface will emit according to eqn. (6.11) so that the total current will be determined essentially by the low work function patches because of the exponential dependence on Φ.

(b) *Photoelectric emission*

An electron with energy below the vacuum level may receive sufficient energy from a photon incident on the surface to be ejected from the crystal. Since the density of occupied states falls off rapidly at the Fermi energy, in the case of a metal, the emitted current exhibits

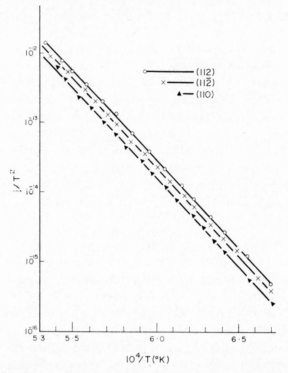

FIG. 6.24. Richardson–Dushman plot for different planes of tungsten. (From F. L. Hughes, Levinstein, and Kaplan, *Phys. Rev.* **113**, 1023 (1959).)

a threshold at a value of the incident photon energy of $h\nu_0 = \Phi$. The emitted current density is dependent on temperature, through the distribution in energy of the electrons in the metal, and on the intensity and frequency ν of the incident radiation as described by the Fowler equation[†]

$$j = BT^2 f\left\{\frac{h(\nu - \nu_0)}{kT}\right\} \tag{6.12}$$

for values of the photon energy near the threshold. B is proportional

[†] For a derivation see R. H. Fowler, *Phys. Rev.* **38**, 45 (1931).

to the incident intensity and is essentially constant for photon energies near the threshold, and the function $f\{h(v-v_0)/kT\}$ can be calculated for a free electron metal. The threshold energy, and hence the work function, can be obtained from a plot of $\ln(j/T^2)$ versus hv/kT obtained either by varying the temperature at fixed incident frequency or by fixing the temperature and varying the frequency.

For a nonuniform surface the work function obtained from the photoelectric method is also strongly weighted toward those patches of low work function.

Recently the energy distribution of emitted photoelectrons from surfaces has been used to provide information on band structures and surface states.

(c) *Field emission*

We have already discussed the phenomenon of field emission of electrons from a metal surface due to the action of a strong electric field. The process is one in which conduction electrons tunnel through a surface barrier which has been lowered and made narrow by the applied field. The effect can be described by the Fowler–Nordheim equation [eqn. (6.1)] and the values of the work function obtained from the slope of a so-called Fowler–Nordheim plot, i.e., a plot of $\ln(I/V^2)$ versus $1/V$ where I is the emitted current and V the applied voltage. Recently the technique has been used most extensively with the geometrical arrangement of the field emission microscope. By collecting the current emitted from a particular plane it is possible to study the variation of Φ with surface orientation in this way. There still appears, however, to be some difficulty associated with determining the correct value of the local field to be used.

6.5.2. *Contact Potential Difference Measurements*

(a) *Kelvin method*

If two different crystals A and B are used as the plates of a capacitor and an external connection is made, the electron Fermi level becomes constant throughout the system and the electrostatic potential differ-

194 PROPERTIES OF CRYSTAL SURFACES

ence existing across the capacitor gap ΔV is equal to the difference in the work functions of the two surfaces[†]

$$\Delta V = \Phi_A - \Phi_B. \qquad (6.13)$$

The corresponding field in the gap is produced by a charge on the capacitor plate surface $Q = C\,\Delta V$, where C is the capacitance. If the capacitance is changed by changing the plate separation charge will flow in the external circuit to keep the Fermi level uniform and hence maintain the same potential difference across the capacitor. The common method of measuring ΔV is to vibrate one of the surfaces at a fixed frequency, detect the current which is produced at this frequency in the circuit, and then apply an external potential difference V_{ext} in series with the capacitor to null the current. The vibrating electrode is then moving in zero electric field and hence ΔV is equal in magnitude to V_{ext} and of opposite sign. If the absolute value of the work function of one of the surfaces is known from an emission type of measurement, the Kelvin method then gives that of the other.

The work function determined in this way for a nonuniform surface represents an average of the various patches in which the contribution from a particular type of patch is weighted according to the fraction of the total surface area which it occupies. The method is particularly well suited to the study of work function changes accompanying adsorption. If one of the surfaces (the reference) is inert to the adsorbing gas, the work function changes of the other can be determined continuously as a function of exposure.

(b) *Retarding potential methods*

This type of measurement is generally most useful for detecting changes in the work function of a particular crystal surface. A low energy electron beam is incident on the specimen surface generally

---

[†] This is strictly true only if the field $\Delta V/d$, where d is the capacitor gap, is sufficiently small that it does not significantly distort the surface charge distributions characteristic of the isolated surfaces. With $\Delta V \sim 1$ V, $d \sim 0.1$ mm, the field is only of the order of 0.1 kV/cm and hence is negligible.

at normal incidence (using, for example, the arrangement of a LEED apparatus (Fig. 6.14)) and the current collected by the specimen is measured as a function of the retarding potential applied to it. The work function change, produced, for example, by adsorption, can then be measured from the displacement of the collected current-retarding potential characteristics. For a meaningful measurement of the work function change the current-retarding potential characteristics have to be accurately parallel, and some difficulties can arise from stray electric and magnetic fields and from diffraction effects within the voltage range which is of interest.

A very elegant modification of the retarding potential method and one which yields an absolute value of the work function has recently been introduced. Electrons from a field emitter A pass through a hole in the anode, are decelerated by a set of grids and finally impinge on the surface of the specimen B which is electrically connected to the emitter. Since in field emission the electrons leaving A originate primarily from the Fermi level, they will arrive at B with an energy which is independent of the emitter work function and which is Φ_B below the potential of the specimen surface so that the work function of B may be determined from a plot of collected current against bias voltage extrapolated to zero current.

(c) *Diode methods*

There are two methods for measuring contact potential differences which depend on determining the current–voltage characteristics of a diode. In the first the diode is operated at very low emission current so that space charge effects are not important. As long as the voltage V_{ext} applied to the anode is sufficient to balance out the contact potential difference ΔV, the current has its saturation value $J_{\max}$. Whenever the direction of the field in the cathode–anode gap is reversed the current falls rapidly according to

$$J/J_{\max} = \exp\{e(\Delta V + V_{\text{ext}})/kT\} \qquad (6.14)$$

for $(\Delta V + V_{\text{ext}}) \leqslant 0$. The applied voltage corresponding to the break in the characteristic thus gives the contact potential difference.

The second method is suitable for detecting work function changes due, for example, to adsorption. In this the diode is operated at much higher saturation currents so that when the accelerating field from the cathode to anode is small, the current is limited by space charge effects. Changes in work function due to adsorption or other causes at the anode can be detected by a shift in the current–voltage charac-

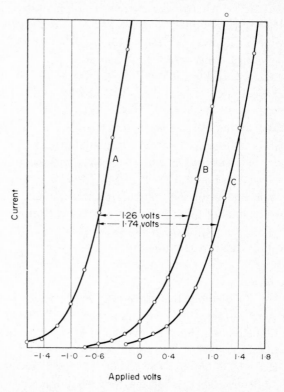

FIG. 6.25. Current–voltage characteristics for a diode arrangement in which one surface (the emitter) is a nominally clean tungsten filament at 2050°K and the other (the collector) a tungsten filament at room temperature. Curve A refers to a nominally clean collector, curve B to a hydrogen-covered collector, and curve C to an oxygen-covered collector. The displacements from A are the contact potential differences produced by the adsorbed layers. (From R. C. L. Bosworth and E. K. Rideal, Studies in contact potentials, *Proc. Roy. Soc.* **A162**, 1 (1937).)

teristic as in the retarding potential method. Figure 6.25 illustrates the effect for the case of hydrogen and oxygen adsorption on polycrystalline tungsten filament surfaces.

Various modifications of these techniques have been used and a very large amount of data on work functions and contact potential differences can be found in the literature. Much of the data applies to polycrystalline ribbons, filaments, or thin films formed by vacuum evaporation, but there is also quite a large amount on single crystal surfaces. A few of these results are listed in Table 6.1. The work function varies with crystallographic orientation for a particular material as shown by the example of tungsten which has been studied more than any other material. It is rather remarkable, however, by how little the work function varies from one metal to another even although there is considerably greater variation in individual inner potentials and Fermi energies (measured from the bottom of the conduction band).

TABLE 6.1. SOME EXPERIMENTAL VALUES OF WORK FUNCTIONS FOR METAL SURFACES[a]

Material	Orientation	Φ (eV)	Method
W	(110)	5.15	Kelvin method
		5.30	Thermionic emission
		5.84	Field emission
	(113)	4.71	Kelvin method
	(100)	4.67	Kelvin method
		4.58	Thermionic emission
	(111)	4.40	Thermionic emission
		4.40	Field emission
Ta	(110)	4.75	Thermionic emission
	(112)	4.35	Saturated diode
Ni	(100)	5.22	Photoelectric emission
Fe	(100)	4.24	Kelvin method
Au	Polycrystalline film	4.28	Kelvin method
Cu	Polycrystalline film	4.51	Kelvin method

[a] Taken from the article by J. C. Rivière (*Solid State Surface Sci.* **1**, 179 (1969)). This article contains a fairly up to date listing of data with critical comments.

6.6. OTHER TECHNIQUES
IN SURFACE STUDIES

In this chapter we have discussed a few of the techniques that have been used fairly extensively in the study of the surface properties of single crystals, mainly metals. There are, of course, a great many other techniques for surface studies, and in fact almost any experimental method in solid state physics or physical chemistry can be adapted for this purpose. Some selected references are given in the bibliography.

Among the most elegant experimental work in surface physics has been that on the electronic properties of the surfaces of elemental and compound semiconductors. Here the main emphasis has been on the properties of electronic surface states and they have been investigated by measuring excess conductance parallel to the surface, differential capacitance of the surface region, surface photovoltages, and photoelectric emission. These are discussed in the texts by Many, Goldstein, and Grover, and by Frankl. It is probable that many of these same techniques could profitably be applied to a range of ionic crystals.

The experimental study of surface atom vibrations is really still in a very early stage of development. Techniques which offer considerable promise in this area include infrared absorption spectroscopy, the Mössbauer effect, and high resolution (~ 10 meV) energy loss studies with very low energy ($\lesssim 10$ eV) electrons. Studies of the scattering of low energy atomic or molecular beams by surfaces may also be useful in investigating surface phonons.

The binding of adsorbed atoms and molecules on surfaces has been studied by a variety of methods including calorimetric measurements, the determination of adsorption isotherms, flash desorption measurements, electron and photon stimulated desorption, and field evaporation. We will discuss some of these methods in Chapter 7.

A technique known as Auger neutralization spectroscopy for the study of the energy bands of solids promises to be extremely useful in surface studies. The quantity measured here is the kinetic energy distribution of electrons ejected from the solid when low energy ions, usually He^+, are incident on the surface. When an ion is incident on the surface an electron may tunnel from an occupied state of the crystal to the ionization level. The energy released in this process of neutralization may be transferred to an electron in another state of the solid which may then be ejected as an Auger electron. The technique is capable of providing information on the energies and densities of surface states, particularly those associated with adsorbed layers.

Other methods include the study of electron spin resonance for following surface reactions, nuclear magnetic resonance for surface atomic diffusion measurements, dielectric constant changes during adsorption of polar molecules for obtaining information on their orientation, and ellipsometry, and vacuum microbalance methods in measuring the rates of surface oxidation and other reactions.

BIBLIOGRAPHY

Surface Preparation

FARNSWORTH, H. E., Atomically clean solid surfaces—Preparation and evaluation, in *The Solid–Gas Interface*, Vol. I (ed. E. A. Flood), Dekker, New York, 1967, chap. 13.

GOLDFINGER, G. (ed.), *Clean Surfaces: Their Preparation for Interfacial Studies*, Dekker, New York, 1970.

REDHEAD, P. A., HOBSON, J. P., and KORNELSON, E. V., *The Physical Basis of Ultrahigh Vacuum*, Chapman & Hall, 1968.

ROBERTS, R. W., and VANDERSLICE, T. A., *Ultrahigh Vacuum and its Applications*, Prentice-Hall, 1963.

Optical Microscopy

BORN, M. and WOLF, E., *Principles of Optics*, Pergamon, 1965.

BRANDON, D., *Modern Techniques in Metallography*, Butterworths, London, 1966.

TOLANSKY, S., *An Introduction to Interferometry*, Longmans, 1955.

TOLANSKY, S., *Multiple Beam Interference Microscopy of Metals*, Academic Press, 1970.

Electron Microscopy

BELK, J. A., and DAVIES, A. L. (ed.), *Electron Microscopy and Microanalysis of Metals*, Elsevier, 1968.

HEIDENREICH, R. D., *Fundamentals of Transmission Electron Microscopy*, Interscience, 1964.

HIRSCH, P. B., HOWIE, A., NICHOLSON, R. B., PASHLEY, D. W., WHELAN, M. J., *Electron Microscopy of Thin Crystals*, Butterworths, London, 1965.

SIEGEL, B. M. (ed.), *Modern Developments in Electron Microscopy*, Academic Press, 1964.

THOMAS, G., *Transmission Electron Microscopy of Metals*, Wiley, New York, 1962.

Scanning Electron Microscopy

KAMMLOTT, G. W., Some aspects of scanning electron microscopy, *Surface Sci.* **25**, 120 (1971).

OATLEY, C. W., NIXON, W. C., PEASE, R. F. W., Scanning electron microscopy, *Advances in Electronics and Electron Physics*, **21**, 181 (1965).

THORNTON, P. R., *Scanning Electron Microscopy*, Chapman & Hall, London, 1968.

Field Electron Emission and Field Ion Microscopy

BRANDON, D., The structure of field-evaporated surfaces, *Surface Sci.* **3**, 1 (1964).

GADZUK, J. W., and PLUMMER, E. W., Field emission energy distributions, *Rev. Modern Phys.* (1972).

GOMER, R., *Field Emission and Field Ionization*, Harvard University Press, 1961.

HOCHMAN, R. F., MÜLLER, E. W., and RALPH, B. (ed.), *Applications of Field Ion Microscopy* in *Physical Metallurgy and Corrosion*, Georgia Tech. Press, Atlanta, 1969.

HREN, J. J. (ed.), Field-ion, field emission microscopy and related topics, *Surface Sci.* **23**, (1) (1970).

MÜLLER, E. W., and TSONG, T. T., *Field Ion Microscopy*, Elsevier, Amsterdam, 1969.

Electron Diffraction from Surfaces

BAUER, E., Reflection electron diffraction *and* Low energy electron diffraction, in *Techniques of Metals Research*, Vol. 11, Wiley, 1969, chaps. 15 and 16.

ESTRUP, P. J., and McRAE, E. G., Surface studies by electron diffraction, *Surface Sci.* **25**, 1 (1971).

LANDER, J. J., Low energy electron diffraction and surface structural chemistry, *Progress in Solid State Chemistry* **2**, 26 (1965).

MAY, J. W., Discovery of surface phases by low energy electron diffraction (LEED), *Advances in Catalysis* **21**, 151 (1970).
PASHLEY, D. W., Recent developments in the study of epitaxy, *Recent Progress in Surface Science* **3**, 23 (1970).
WOOD, E. A., Vocabulary of surface crystallography, *J. Appl. Phys.* **35**, 1306 (1964).

Surface Composition Analysis

CHANG, C. C., Auger electron spectroscopy, *Surface Sci.* **25**, 53 (1971).
HARRIS, L. A., Analysis of materials by electron-excited Auger electrons, *J. Appl. Phys.* **39**, 1419 (1968).
PARK, R. L., HOUSTON, J. E., and SCHREINER, D. G., A soft X-ray appearance potential spectrometer for the analysis of solid surfaces, *Rev. Sci. Instr.* **41**, 1810 (1970).
RIVIÈRE, J. C., Characteristic Auger electron emission as a tool for the analysis of surface composition, *Physics Bulletin* **20**, 85 (1969).
SIEGBAHN, K., *et al.*, Atomic, molecular and solid state structure studied by means of electron spectroscopy, *Nova Acta Regiae Soc. Sci. Upsaliensis*, Ser. 4, **20** (1967).
SMITH, D. P., Analysis of surface composition with low-energy backscattered ions, *Surface Sci.* **25**, 171 (1971).
TAYLOR, N. J., The technique of Auger spectroscopy in surface analysis, in *Techniques of Metals Research*, Vol. VII, Interscience (1971).

Work Function Measurements

HOLSCHER, A. A., A field emission retarding potential method for measuring work functions, *Surface Sci.* **4**, 89 (1966).
RIVIÈRE, J. C., Work function: Measurements and results, *Solid State Surface Sci.* **1**, 179, Dekker, New York (1969).

Other Techniques

FLOOD, E. A. (ed.), *The Solid–Gas Interface*, Vol. II, Dekker, 1967.
FRANKL, D. R., *Electrical Properties of Semi-conductor Surfaces*, Pergamon, 1967.
HAGSTRUM, H. D., and BECKER, G. E., Orbital energy spectra of electrons in chemisorption bonds: O, S, Se on Ni (100), *J. Chem. Phys.* **54**, 1015 (1971).
HAIR, M. L., *Infrared Spectroscopy in Surface Chemistry*, Dekker, New York, 1967.
LITTLE, L. H., *The Infrared Spectra of Adsorbed Molecules*, Academic Press, London, 1966.
SELWOOD, P. W., *Adsorption and Collective Paramagnetism*, Academic Press, New York, 1962.
Ellipsometry in the Measurement of Surfaces and Thin Films, US National Bureau of Standards, Pub. No. 256, 1964.

CHAPTER 7

ATOMIC PROCESSES
AT SURFACES

7.1. INTRODUCTION

In this chapter we discuss two processes which are involved in a number of reactions at crystal surfaces—surface atomic diffusion and adsorption. These constitute important steps in such phenomena as oxidation, heterogeneous catalysis, condensation and evaporation, and the nucleation and epitaxial growth of thin films. For example, in producing a single crystal film by condensation from the vapor, individual atoms will first be adsorbed on the surface, diffuse around among the various adsorption sites, interact with each other either to produce an adsorbed structure with long range order, or aggregate into island nuclei from which the final crystal is generated.

A variety of different mechanisms may be involved in the motion of surface atoms and there are several ways in which surface diffusion may differ significantly from diffusion in the crystal interior. In some cases the motion may best be described in terms of a two-dimensional gas model, while in other systems a model of diffusion between equilibrium sites is more appropriate. Surface diffusion plays an important part in a number of mass transport or shape change processes such as sintering which occur with crystals at elevated temperatures. We will discuss the basic aspects of mass transport due to capillarity in terms of the development of rather simple morphologies.

Research work on the phenomenon of adsorption has been directed at determining the nature of the binding to the surface, the equilibrium between adsorbed and gaseous states, the interaction among adsorbed molecules (which often leads to the formation of ordered layers), and the rates of adsorption and desorption. Although a great deal of data has been obtained, many aspects of adsorption are not yet fully understood. We will consider some of the ideas which have emerged from this work and discuss a few simple models.

7.2. SURFACE ATOMIC DIFFUSION

In the immediate vicinity of the free surface of a solid or liquid the mobility of the atoms may differ significantly from those in the bulk of the material. This can be described in terms of a surface diffusivity which may be defined through the use of surface excess quantities as described in Chapter 1. The normal bulk diffusivity (or diffusion coefficient) of a particular component is defined through Fick's phenomenological diffusion equation $\mathbf{j} = -D\,\mathrm{grad}\,c$, where $\mathbf{j}$ is the flux density (atoms/(cm²-sec)) of the diffusing species, c its local concentration (atoms/cm³) and D the appropriate diffusivity coefficient (cm²/sec) (D may, of course, be a function of concentration). We also define the surface diffusivity using Fick's law in the following way.

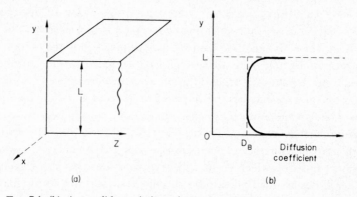

FIG. 7.1. (b) A possible variation of atomic self-diffusivity with position across the slab of crystal as shown in (a). The diffusivity has the value D_B in the bulk of the crystal and is assumed to increase in the surface regions. The surface diffusivity may be defined in terms of the excess surface flux and concentration gradient.

Consider the hypothetical solid shown in Fig. 7.1 in which the concentration gradients are solely in the x-direction. Then the integrated flux J of a particular species along the solid in the x-direction per unit length in the z-direction is

$$J = -\left(\frac{dc}{dx}\right) \int_0^L D(y)\, dy, \qquad (7.1)$$

where, as indicated, the diffusion coefficient is in general a function of y. The corresponding integrated flux J^B which would obtain if the solid were uniform to the planes $y = 0$ and $y = L$ (the dividing surfaces) is given by

$$J^B = -D_B \left(\frac{dc}{dx}\right) L, \qquad (7.2)$$

where D_B is the diffusion coefficient in the homogeneous solid. The surface excess flux J^s, per unit length in the y-direction, at each of the two bounding surfaces is then defined (following eqn. (1.1)) as

$$J^S = \tfrac{1}{2}(J - J^B), \qquad (7.3)$$

i.e.

$$J^S = -\frac{1}{2}\left(\frac{dc}{dx}\right) \int_0^L \{D(y) - D_B\}\, dy. \qquad (7.4)$$

A surface diffusion coefficient D_S with units of cm²/sec may be introduced by setting the integral in eqn. (7.4) equal to $2D_S\delta$, i.e. the excess surface flux is the same as would be obtained from a layer of thickness δ and in which the diffusion coefficient is uniform and of magnitude D_S. The choice of the parameter δ is arbitrary, but it is almost invariably taken (usually implicitly) to be one atomic plane separation in the direction of the surface normal. We should note that the surface diffusivity defined in this way could, of course, be negative corresponding to a negative excess flux but that surface contributions to measured fluxes will normally only be measurable when D_S is considerably greater than D_B. Experiments suggest that D_S is usually greater than D_B, but there is no *a priori* reason to assume that this should always be so. With ionic crystals of the type discussed in section 4.4 there will be a contribution not only from the topmost layer but also from the space charge region near the surface. With metallic crystals, on the other hand, the strong shielding effects of the conduction electrons confine differences in defect concentrations and mobilities to the first few atomic layers.

Bulk self-diffusion of atoms in crystals usually involves point defects. There is a considerable body of evidence to suggest that in most face-centered cubic crystals atom motion involves vacant lattice sites while in other cases, notably impurity diffusion in body-centered cubic materials, the mobile atoms move between interstitial sites. Other possible mechanisms of atom motion include site exchange and the various modifications thereof. Each of these has its analogue in the case of surface self-diffusion. If we confine our attention for the moment to the situation where all but the topmost layer has bulk defect densities and mobilities, some possible surface mechanisms can be seen with reference to Fig. 7.2: vacancy diffusion within the nearly

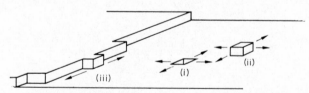

Fig. 7.2. Various possible mechanisms of diffusion on a crystalline surface: (i) surface vacancy diffusion, (ii) adatom diffusion, (iii) diffusion along monatomic steps.

complete layer, self-adsorbed atom diffusion on the nearly complete layer, and diffusion of atoms along steps. The last of these may be compared to the process of diffusion along dislocation cores within crystals. With surfaces which are not close to a singular orientation, the distinction among these various modes of transport will become somewhat less clear and with high index orientations and high temperatures it may not be sensible to distinguish, for example, vacancy and adsorbed atom contributions. However, at relatively low temperatures where there is little disorder and for singular or vicinal surfaces, we can within the one layer model express the surface self-diffusion coefficient D_s as

$$D_s = \sum \frac{n_i}{M} D_i, \tag{7.5}$$

where n_i and M are respectively the number of defects of type i and the number of sites per unit area, and D_i are the corresponding diffusion coefficients of the defects of type i. If the concentrations (n_i/M) of the various defects are the values appropriate to equilibrium, then the diffusivity given by eqn. (7.5) is an intrinsic property of that particular surface, and we may refer to it as the *intrinsic surface diffusion coefficient*. In an analogous way, *extrinsic diffusion coefficient* will refer to a situation where surface diffusion takes place under conditions where the defect concentrations are changed from their equilibrium values by some external means, e.g. by condensation from the vapor, ion bombardment, etc. From the point of view of learning about the mechanism of surface diffusion in particular cases it would be desirable to determine both the intrinsic diffusivity and the extrinsic value under well-specified conditions. Usually only the total quantity D_s is obtained from experiments, but in a few special cases the diffusivities of self-adsorbed atoms have been measured.

At present the mechanism of surface self-diffusion has not been established with certainty for any particular surface. Nevertheless, it is of interest to consider some aspects of the diffusion of atoms on surfaces particularly in so far as they differ from normal diffusion processes within the bulk of crystals. We will consider as an example the process of surface self-diffusion via the mechanism of self-adsorbed atoms. The diffusion coefficient is then a product of the concentration of self-adsorbed atoms and their diffusivity. The concentration is given by

$$\frac{n_a}{M'} = \exp\left(-\frac{\Delta G_0}{kT}\right) \tag{7.6}$$

(see section 4.3.2), and the potential energy variation as seen by a

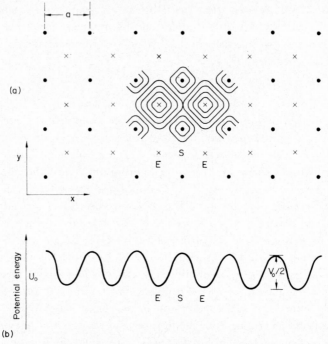

FIG. 7.3. (a) Arrangement of occupied lattice sites (●) and available sites (×) for self-adsorbed atoms on a (100) surface. The contours refer to the potential energy for an adsorbed atom on this type of surface. The equilibrium sites are denoted by E. In (b) the variation of potential energy with position along the line joining neighboring equilibrium sites is represented as a sinusoidal energy variation with a maximum of $V_0/2$ at the saddle points S relative to the equilibrium sites. The magnitude of this diffusion barrier will depend on the particular type of atom which is diffusing.

self-adsorbed atom on the surface may be envisioned as in Fig. 7.3. The potential energy of the adsorbed atom with variation in x and y will be doubly periodic on the singular portions. Treating the variation as being purely harmonic, the potential energy of an adatom on a (100) surface of a cubic crystal may be written as

$$U(x, y) = U_0 + \frac{V_0}{2}\left(1 - \cos\frac{2\pi x}{a}\right) + \frac{V_0}{2}\left(1 - \cos\frac{2\pi y}{a}\right), \quad (7.7)$$

where U_0 is the potential energy of an atom in an equilibrium adsorption site relative to that in the gas phase (taken as the reference zero of potential energy) and $V_0/2$ is the height of the potential barriers which must be surpassed in moving from the equilibrium positions E through the saddle configurations S to the adjacent equilibrium sites. A number of features of the motion of such adsorbed atoms are worth noting. If the barrier height is considerably less than kT, i.e. if $V_0/2 \ll kT$, the self-adsorbed atoms will behave like a two-dimensional gas, the diffusivity being determined by their root mean square velocity and mean free path. If $V_0/2 \gg kT$, the self-adsorbed atoms will spend most of their time in localized vibrational states in the equilibrium sites and their diffusive motion will consist of transitions between equivalent sites. However, even in the latter case there may be considerable difference from motion of atoms in the bulk of crystals. Once an adsorbed atom receives a sufficiently large component of momentum to surmount the potential energy barrier it may travel a significant number of lattice spacings before returning to a localized vibrational state by coupling its excess energy back to the lattice. This might be expected since it would involve very little cooperative motion of the surface atoms. On the other hand, within the crystal the situation is such that diffusive jump distances of more than one lattice spacing are extremely unlikely. This is quite clear for the case of diffusion via vacancies, and even the interstitial diffusion mechanism requires a considerable amount of cooperative motion of the atoms on the lattice sites. In any case the defect diffusivity (adsorbed atom) on the (100) surface may be written as

$$D_a = \frac{1}{4} \frac{l^2}{\tau}, \qquad (7.8)$$

where l is the mean jump distance and τ is the mean time that an atom spends in the localized positions before jumping.

According to the so-called "equilibrium" theory of reaction rates developed principally by Eyring,[†] the mean jump frequency may be

† S. Glasstone, K. J. Laidler and H. Eyring, *The Theory of Rate Processes*, Mc Graw-Hill, New York (1941).

written as

$$\frac{1}{\tau} = \frac{kT}{h} \frac{M'^*}{M'} \left(\frac{q_z^* q_y^*}{q_x q_y q_z}\right) \exp\left(-\frac{V_0}{2kT}\right), \tag{7.9}$$

where M'^* and M' are the numbers of saddle points and of adsorption sites per unit area respectively, q_x, q_y, and q_z the contributions to the partition function of an atom in the adsorbed state corresponding to three (assumed independent) degrees of freedom, and q_y^* and q_z^* the corresponding quantities for the saddle configuration; the degree of freedom corresponding to the direction of motion over the saddle configuration has been taken to be one of translation since the potential energy has negative curvature about this point along the direction of motion or reaction coordinate as it is sometimes called. For the general case, when V_0 may be comparable to kT, the evaluation of the partition functions is rather difficult since the motion is neither pure translation nor pure vibration. It may be referred to as *hindered translation*. The motion in the z-direction, i.e. normal to the surface, may be taken to be pure vibration (in the adsorption well), while for the xy motion it can be shown[†] that the product of the partition functions may be written, for the potential of eqn. (7.7), as

$$q_x q_y = \frac{M' 2\pi U' \exp(-2U' - KU') I_0^2(U')}{\{1 - \exp(-KU')\}^2},$$

where $U' = V_0/2kT$, $K = (2h^2/ma^2 V_0)^{1/2}$ with m the mass of the adsorbed atom, and $I_0(U')$ is a modified Bessel function of the first kind whose full form need not concern us here. Similarly, the saddle configuration can be taken to correspond to the situation of one degree of "hindered translation" and one degree of vibration (along the z-direction). Thus for the doubly harmonic potential of eqn. (7.7),

$$\frac{1}{\tau} = \frac{kT}{h} \frac{M'^*}{M'} \frac{\{1 - \exp(-KU')\}}{(2\pi M' U')^{1/2} \exp(-U' - \frac{1}{2}KU') I_0(U')} \exp\left(-\frac{V_0}{2kT}\right). \tag{7.10}$$

[†] See T. Hill, *An Introduction to Statistical Thermodynamics*, Addison-Wesley, 1960, p. 172.

In the two limiting cases eqn. (7.10) can be simplified significantly. For $V_0 \gg 2kT$, the localized adsorption limit,

$$\frac{1}{\tau} = 2v_x \exp\left(-\frac{V_0}{2kT}\right),$$

where v_x is the frequency of vibration in the equilibrium site along the x-direction, parallel to the surface. Hence

$$D_a = \frac{1}{2} l^2 v_x \exp\left(-\frac{V_0}{2kT}\right), \tag{7.11}$$

the normal type of formula for diffusion between equilibrium sites. On the other hand, when $V_0 \ll 2kT$, the free translation limit, we obtain, using eqn. (7.10) or directly from eqns. (7.8) and (7.9),

$$D_a = \frac{1}{4}\left(\frac{kT}{2\pi m}\right)^{1/2} l, \tag{7.12}$$

the diffusion coefficient of a two-dimensional gas. It appears that only the localized adsorption model has been supported by experiment for the case of self-diffusion on crystals. Some elegant experiments in the field ion microscope have shed some light on this question. Tungsten atoms were deposited at $\sim 20°$K on to a tungsten field ion microscope specimen (prepared by field evaporation) so that some of the deposited atoms could be identified in adsorption sites on the more closely packed planes. The displacements of these adsorbed atoms following short periods of heating to temperatures around room temperature were then observed and adsorbed atom diffusivities D_a calculated from the expression

$$\langle R^2 \rangle = 2D_a t, \tag{7.13}$$

where $\langle R^2 \rangle$ is the observed mean square displacement. The results showed the diffusivity to be significantly orientation dependent with considerable anisotropy on particular planes. As shown in Table 7.1,

TABLE 7.1. DATA ON THE DIFFUSION OF ADATOMS ON TUNGSTEN SURFACES FROM FIELD ION MICROSCOPY STUDIES

	Ta	Mo	W	Re	Ir	Pt
ΔH_a^m (110) W D_{a_0}	0.78 4.4×10^{-2}	— —	0.87 2.1×10^{-3}	1.04 1.5×10^{-2}	0.78 8.9×10^{-5}	~ 0.6 $\sim 10^{-4}$
ΔH_a^m (211) W D_{a_0}	0.49 0.9×10^{-7}	0.56 2.4×10^{-6}	0.57 3.8×10^{-7}	0.88 1.1×10^{-2}	0.58 2.7×10^{-5}	— —
ΔH_a^m (321) W D_{a_0}	0.67 1.9×10^{-5}	— —	0.84 1.2×10^{-3}	0.88 4.8×10^{-4}	— —	— —

ΔH_a^m in eV. D_{a_0} in cm²/sec.

The diffusion coefficient for adsorbed atoms has been expressed as $D_a = D_{a_0} \exp(-\Delta H_a^m/kT)$ where D_{a_0} is the pre-exponential or frequency factor and ΔH_a^m the enthalpy change associated with its motion. The latter will essentially be the barrier height $V_0/2$ in eqn. (7.5).

(From data given by D. W. Bassett and M. J. Parsley, *J. Phys. D.: Appl. Phys.* **3,** 707 (1970).)

the activation energies obtained are in the range 0.5–1 eV, in this particular case suggesting that even at temperatures approaching the melting point individual adsorbed atoms would exist mostly in vibrational states on the surface. An interesting effect observed in these experiments was that the self-adsorbed atoms were apparently reflected from the steps at the edges of the low index portions with a reflection probability greater than $\frac{1}{2}$, suggesting that the potential energy across the step may be as indicated in Fig. 7.4.

Apart from the results of these field ion microscope experiments there are no experimental data on the energies associated with surface defect formation or motion. In most experiments only a total diffusivity has been measured, and it has generally not been possible to identify unambiguously the mechanism of diffusion.

Most of the recent data on surface self-diffusion have been obtained from "mass transport" techniques in which rates of change in shape of crystal surfaces are measured at elevated temperatures. Apart from

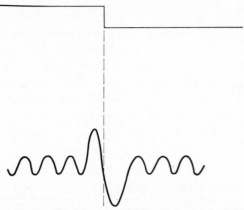

FIG. 7.4. Possible potential energy variation of an adsorbed atom on crossing a monatomic step. For such a variation the adsorbed atom has a greater probability for reflection than for transmission.

providing diffusion data, these processes are of interest in connection with the understanding of a number of phenomena such as sintering, creep, precipitate interaction, and coarsening (aging). In the next section we shall outline the theory of mass transport in crystals due to capillarity and consider a few of the measurements that can be made. Only a limited amount of data on surface diffusion has been obtained by techniques using radioactive tracers although this would appear to be an obvious extension of a technique which has been very successful in bulk diffusion studies. There are special difficulties in the case of surface diffusion measurements associated with the simultaneous diffusion of the isotope into the crystal and with losses from evaporation.

7.3. MASS TRANSPORT DUE TO CAPILLARITY

We will consider the change in shape of crystal surfaces due to the tendency to minimize the surface contribution to the total free energy. In section 1.4 we stated the criteria for determining the equilibrium morphologies of crystals; we now consider the mechanisms by which these equilibrium morphologies can be reached. For simplicity we shall restrict our attention to surfaces of one-component crystals which are nearly planar and consider their development with time. Although these are somewhat artificial conditions they will allow us to cover the main features of the theory with little mathematical complexity.

There are a number of mechanisms which could lead to a shape change of a crystal surface, viz. diffusion, evaporation–condensation processes, and plastic flow. We consider first the transport by diffusion in a system which is only slightly removed from equilibrium; the diffusional contribution may then be expressed in terms of linear flux–force relationships in which the flux density of atoms $\mathbf{j}$ is related to the gradients in chemical potential by

$$\mathbf{j} = -\frac{Dc}{kT} \operatorname{grad} (\mu - \mu_v), \tag{7.14}$$

where c is the atom concentration, D the atomic diffusion coefficient[†] and μ and μ_v the atom and vacancy chemical potential respectively. The method of solution of particular cases is to make the assumption that the flow is divergenceless except at the surface, i.e.

$$\operatorname{div} \{Dc \operatorname{grad} (\mu - \mu_v)\} = 0. \tag{7.15}$$

With some assumption about the variation of diffusion coefficient with depth the solution of this equation, with the boundary conditions that $(\mu - \mu_v)$ satisfy relationships of the form of eqn. (1.26) at the surface, gives $(\mu - \mu_v)$ as a function of position. From this the resulting flux and hence the net flux into any element of the surface may be evaluated so that the evolution of the surface can be described. It has generally been assumed that the diffusivity is in fact constant except for the last atomic layer of the crystal. In this case eqn. (7.15) may be approximated by the Laplace equation in $(\mu - \mu_v)$. Of special interest is the evolution of a sinusoidal surface since any arbitrary

[†] The diffusion coefficient may differ from the diffusivity determined by a tracer technique. For bulk diffusion in a f.c.c. crystal via a vacancy mechanism, for example, the tracer diffusivity is less than the mass transport value by a factor of about 0.78, the correlation factor (see, for example, P. G. Shewmon, *Diffusion in Solids*, McGraw-Hill, 1963).

shape can be built up of an appropriate combination of sine waves
and also since such surfaces can in fact be produced, and measurements of their decay with time have been used in the evaluation of
diffusion coefficients.[†]

Consider the effect of bulk diffusion on a sinusoidal surface[‡] described by

$$z_s = a \sin Kx \tag{7.16}$$

with $K = 2\pi/\lambda$, where λ is the repeat distance of the sinusoidal corru-

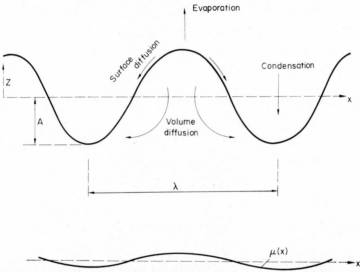

FIG. 7.5. Schematic of a sinusoidal corrugation on a surface. The possible
modes by which the surface can approach planarity are indicated on the
figure. For sufficiently small wavelengths λ the process of surface diffusion
will be dominant; for the metals which have been studied this is generally
true below about 20 μ. The lower curve indicates the corresponding variation of the chemical potential of atoms immediately beneath the surface.

[†] Similar techniques have been used for the study of mass transport processes
at metal–ceramic interfaces (see, for example, *Surfaces and Interfaces I*, p. 325,
1967).
[‡] The description given here is derived mainly from work of W. W. Mullins
reviewed in *Metal Surfaces*, American Society for Metals, Cleveland, 1963,
chap. 2.

15*

gation Fig. 7.5. The boundary condition on the chemical potentials according to section 1.6 is at $z_s = z$,

$$(\mu - \mu_v) = (\mu - \mu_v)_0 + \gamma \Omega_0 \left\{ \frac{1}{R_1} \left(1 + \frac{\gamma''}{\gamma} \right) \right\}, \qquad (7.17)$$

where $\gamma'' = \partial^2 \gamma / \partial n_x^2$ and the curvature

$$\frac{1}{R_1} = -\frac{\partial^2 z_s / \partial x^2}{\{1 + (\partial z_s / \partial x)^2\}^{3/2}} \approx -\frac{\partial^2 z_s}{\partial x^2} \qquad (7.18)$$

provided the surface is everywhere of small slope. If we assume that the vacancies are in equilibrium everywhere so that $\mu_v = 0$, the boundary conditions become

$$\mu = \mu_0 + \Omega_0(\gamma + \gamma'') aK^2 \sin Kx \quad \text{at} \quad z = z_s$$

and

$$\mu = \mu_0 \quad \text{at} \quad z = -\infty. \qquad (7.19)$$

The solution of $\nabla^2 \mu = 0$ for these conditions is readily shown to be

$$\mu = \mu_0 + \Omega_0(\gamma + \gamma'') aK^2 (\sin Kx) \exp (Kz) \quad \text{for} \quad z \leqslant z_s, \qquad (7.20)$$

and the flux of atoms at any point $\mathbf{j}(x, z)$ is given by

$$\mathbf{j}(x, z) = -\frac{D_B c}{kT} \operatorname{grad} \{\Omega_0(\gamma + \gamma'') aK^2 (\sin Kx) \exp (Kz)\}. \qquad (7.21)$$

To compute the evolution of the profile due to bulk diffusion we need only compute the normal component j_z of the flux to any element of the surface. With the assumption that the normal to the actual surface deviates only slightly from its average direction, this is approximately given by

$$j_z = -\frac{D_B(\gamma + \gamma'')}{kT} aK^3 \sin Kx \quad \text{at} \quad z = z_s, \qquad (7.22)$$

where we have set $c = \Omega_0^{-1}$.

Thus

$$\frac{\partial z_s}{\partial t} = \Omega_0(j_z)_{z=z_s} = -\frac{D_B\Omega_0(\gamma+\gamma'')}{kT} K^3 z_s. \qquad (7.23)$$

The differential equation describing the change in shape of the surface with time due to bulk diffusion alone is then

$$\frac{\partial z_s}{\partial t} = -CK^3 z_s \qquad (7.24)$$

with solution

$$z_s(x, t) = z_s(x, 0) \exp(-CK^3 t), \qquad (7.25)$$

where $C = [D_B\Omega_0(\gamma+\gamma'')]/kT$. Thus the surface maintains its sinusoidal shape and the amplitude decays exponentially with time. (In asserting that it decays we are in effect assuming that $(\gamma+\gamma'')$ is positive, whereas if $(\gamma+\gamma'')$ were negative the undulation would, of course, grow with time; $(\gamma+\gamma'')$ is in fact positive in those cases where no faceting is observed). There are, of course, other mechanisms which will contribute to the change in shape of surfaces, and these are indicated in Fig. 7.5 for the special case of the sinusoidal corrugation. The importance of the individual transport mechanisms depends upon the characteristic dimensions of the surface under study and, as will be shown below for the sinusoidal case, the process of surface diffusion will usually be most important at small wavelengths while volume diffusion evaporation–condensation or even plastic flow will dominate for large wavelengths.

The flow of material due to surface diffusion (i.e. the excess surface flux) is given as above by

$$\mathbf{J}^s = -\frac{D_s c_s}{kT} \operatorname{grad} (\mu - \mu_v). \qquad (7.26)$$

This equation could be taken to define the product $D_s c_s$, where D_s will in general be a second rank tensor. However, according to the convention introduced earlier, c_s is taken to be the number of atoms per unit area in one monolayer.

At any point on the surface the net accumulation of material is given by

$$-\operatorname{div} \mathbf{J}^s = \frac{D_s c_s}{kT} \nabla^2(\mu - \mu_v), \qquad (7.27)$$

so that the rate at which any element moves normal to itself due to surface diffusion alone is

$$\frac{\partial z_s}{\partial t} = -(\operatorname{div} \mathbf{J}^s)\Omega_0 = -\frac{D_s c_s}{kT} \Omega_0(\gamma + \gamma'') aK^4(\sin Kx),$$

i.e. $\quad \dfrac{\partial z_s}{\partial t} = -BK^4 z_s \qquad (7.28)$

with solution

$$z_s(x, t) = z_s(x, 0) \exp(-BK^4 t). \qquad (7.29)$$

Thus in this case also the sinusoidal shape is preserved and the amplitude should decay exponentially with time, the decay constant being inversely proportional to the fourth power of the wavelength.

The other important mode of mass transport in crystals is apparently that involving evaporation and condensation processes. When a crystal is in contact with its vapor a dynamic equilibrium will eventually be set up in which the rate of condensation is exactly balanced by the rate of evaporation. However, due to capillarity the equilibrium pressure may vary from point to point on the surface, and hence some regions will experience net evaporation while others will experience net condensation. If p is the pressure of the vapor the number of atoms condensing per unit area per unit time is, according to the kinetic theory of gases,

$$\alpha \frac{p}{(2\pi mkT)^{1/2}}, \qquad (7.30)$$

where α is a condensation coefficient ($\leqslant 1$) and m the mass of an atom of the solid. For an element of surface of curvature K the vapor pressure p_K that would be in equilibrium with this element is related to that over a flat surface, i.e. p_0, by

$$K(\gamma+\gamma'')\Omega_0 = kT \ln (p_K/p_0) \qquad (7.31)$$

from eqns. (1.21) and (1.27) (pp. 25, 28). If the vapor is sufficiently dilute that the mean free path is large compared with the spacing of surface undulations, the rate of condensation will be independent of curvature and will be given by eqn. (7.30) with p set equal to p_0. Thus the net rate of condensation on an element of curvature K is

$$\alpha \frac{(p_0-p_K)}{(2\pi mkT)^{1/2}} \approx - \frac{\alpha p_0}{(2\pi m)^{1/2}(kT)^{3/2}} \Omega_0(\gamma+\gamma'')K, \qquad (7.32)$$

and hence the motion of the surface due to this process is, for the sinusoidal case,

$$\frac{\partial z_s}{\partial t} = -\frac{\alpha p_0 \Omega_0^2(\gamma+\gamma'')}{(2\pi m)^{1/2} (kT)^{3/2}} K^2 z_s, \qquad (7.33)$$

i.e.

$$\frac{\partial z_s}{\partial t} = -AK^2 z_s, \qquad (7.34)$$

which has the solution

$$z_s(x, t) = z_s(x, 0) \exp(-AK^2 t). \qquad (7.35)$$

Because each of the mechanisms considered leads to an exponential decay which preserves the sinusoidal profile, they can be combined to give as the solution for the decay of a sine wave due to the simultaneous action of volume diffusion, surface diffusion, and evaporation–condensation,[†]

$$z_s(x, t) = z_s(x, 0) \exp\{-(AK^2+CK^3+BK^4)t\}. \qquad (7.36)$$

Surface diffusion will usually dominate at sufficiently small values of

---

† The symbols C, B, and A are used here to correspond with current practice in the research literature on this topic.

the corrugation period λ. For the metals in which such relaxations have been observed experimentally the surface contribution appears to be dominant for values of $\lambda \lesssim 20$ μ at temperatures up to the melting point.

Having obtained the solution for a single sine wave the formal solution for any arbitrary surface shape (involving only small slopes) can be obtained by suitably combining a series of sinusoidal components. Thus the initial shape may be written as

$$z_s(x, 0) = \int_{-\infty}^{\infty} a(K, 0) \exp (iKx) \, dK,$$

where

$$a(K, 0) = \frac{1}{2\pi} \int_{-\infty}^{\infty} z_s(x, 0) \exp (-iKx) \, dx$$

is the amplitude of the component of wavelength $\lambda \, (= 2\pi/K)$. The development with time is then, corresponding to eqn. (7.36),

$$z_s(x, t) = \int_{-\infty}^{\infty} a(K, 0) \exp (iKx) \exp \{-(AK^2 + CK^3 + BK^4)t\} \, dK.$$

$$(7.37)$$

Because of the strong dependence of the damping rate on wavelength for each of the mechanisms, it is clear that the short wavelength contributions to the corrugation will rapidly disappear, leaving only the lower wavelength contributions which involve smaller curvatures. These features are illustrated in Fig. 7.6 which shows the flattening of a periodic surface of a single crystal of nickel.

There are two other possible modes of mass transport near surfaces: diffusion in the vapor phase and viscous or plastic flow. Vapor phase diffusion will be important when the mean free path of atoms in the vapor is small compared to the characteristic distance associated with the surface roughness. These circumstances may arise with sufficiently concentrated vapor or when an inert gas is present and con-

$[1\bar{1}0]$

$[110]$

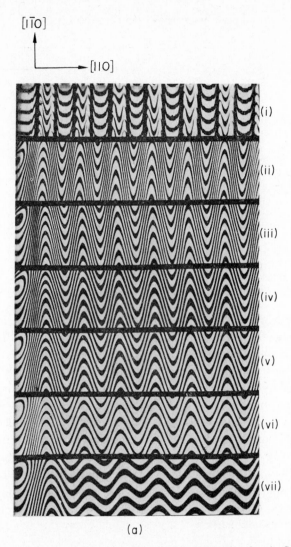

(i)

(ii)

(iii)

(iv)

(v)

(vi)

(vii)

(a)

FIG. 7.6(a). Series of two-beam interference micrographs showing the flattening of a periodic corrugation on a single crystal surface of nickel in vacuum at $\sim 1490°$K. The initial corrugation (i) is produced by a photoetching technique. With annealing the higher Fourier components of this shape disappear rapidly to form the nearly harmonic surface of (ii). The annealing times are (ii) 44 hr, (iii) 59 hr, (iv) 83 hr, (v) 97 hr, (vi) 120 hr, (vii) 196 hr. The amplitude decays exponentially with time as indicated in Fig. 7.6(b).

222   PROPERTIES OF CRYSTAL SURFACES

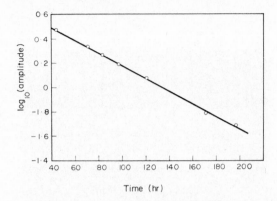

FIG. 7.6(b). In this particular case the main mechanism causing flattening
is that of surface diffusion. (From P. S. Maiya and J. M. Blakely, Surface
self-diffusion and surface energy of nickel, *J. Appl. Phys.* **38**, 698 (1967).)

centration gradients will then be set up in the gas phase leading to
diffusional flow. The contribution from vapor phase diffusion will
have the same dimensional dependence as for diffusion in the crystal
and the corresponding decay constant will be $C_g K^3$, where $C_g =
c_g D_g \Omega_0^2 (\gamma + \gamma'')/kT$ and C_g and D_g are respectively the number of
molecules (from the solid) per unit volume in the gas phase and their
diffusivity. The initiation of substantial plastic flow in crystals by dislocation motion as a result of the tendency to decrease surface free energy
seems to be very unlikely except, perhaps, for extremely high curvatures. Only in the latter case are sufficiently high stresses present to
lead to significant dislocation motion. However, in glassy materials
the process of viscous flow may be significant, and it can be shown
that a sinusoidal surface of a glass should decay due to viscous flow
according to

$$\frac{\partial z_s}{\partial t} = -FKz_s, \qquad (7.38)$$

where $F = \gamma/2\mu$ with μ the coefficient of viscosity. Using the Ein-
stein–Stokes equation to relate the diffusion coefficient D and the vis-

cosity (i.e. $D\mu = kT/3\pi d$, where d is the molecular diameter), it is readily shown that viscous flow will generally be more significant than volume diffusion in effecting mass transport in glassy materials.

To this point we have restricted our attention to the smoothing of sinusoidal corrugations on surfaces since this served to illustrate the principles involved without mathematical complexities. There are, of course, other cases which are of interest in connection with diffusion measurements. One of these is the development of a groove at the line of intersection of a grain boundary with a free surface (Fig. 7.7). The angle 2β of the groove is determined by the specific free energies of the grain boundary and the surface (section 1.5); the increase in width and depth with time is due to the transport of material from the high curvature region near the root of the groove to flatter parts of the surface. When the groove formation is by a diffusional mechanism, the morphology is as indicated in Fig. 7.7(a) with maxima on either side, while the evaporation–condensation process produces the profile indicated in Fig.7.7(b). For the three main mass transport processes in crystalline materials the characteristics of the groove can be shown[†] to be related to the mass transport parameters in the following way.

Surface diffusion only:

$$w = 4.6(Bt)^{1/4}$$
$$d = 0.97m(Bt)^{1/4}, \quad \text{where} \quad m = \tan \beta. \tag{7.39}$$

Volume diffusion only:

$$w = 5.0(Ct)^{1/3},$$
$$d = 1.01m(Ct)^{1/3}. \tag{7.40}$$

Evaporation–condensation only:

$$d = 1.13m(At)^{1/2}. \tag{7.41}$$

It is clear that measurement of the development of the width and depth

---

† See footnote †, p. 215.

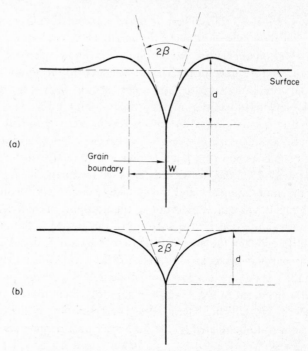

FIG. 7.7. Possible profiles produced at the intersection of a grain boundary with a free surface. The equilibrium dihedral angle 2β is given (approximately) by $\cos \beta = \gamma_B/2\gamma_s$, where γ_B is the grain boundary tension. In (a) humps appear along each side of the groove, the maxima being separated by a distance W; such shapes are characteristically produced when the main mass transport process is that of volume or surface diffusion. In (b) no maxima occur in the profile; this is the characteristic shape of a groove produced by an evaporation and condensation process. For high angle grain boundaries in f.c.c. metals the angle 2β is actually $\sim 160°$.

of grain boundary grooves as indicated in Fig. 7.8 can provide information on surface or bulk diffusion coefficients. Other mass transport phenomena used in diffusion measurements include the smoothing of individual scratches on surfaces, the sintering of spherical crystals and the rate of blunting of needle-type specimens used in the field emission microscope. The principles involved in describing these various phenomena are the same as those for the sine wave treated here although the details may be rather complicated. A survey of the surface diffusion results obtained from experiments of this type are presented in Table 7.2. The table gives values of the frequency factor D_0 and the activation energy Q_s for surface diffusion when the surface diffusivity is expressed in the conventional form $D_s = D_0 \exp (-Q_s/kT)$. Although there are still some significant disagreements among the available data it seems to be clear that the surface diffu-

FIG. 7.8(a). Interferogram showing the profile produced by annealing at the intersection of a grain boundary with a free surface in copper. (From N. A. Gjostein in *Metal Surfaces* (ed. N. A. Gjostein and W. D. Robertson), American Society for Metals, Cleveland, 1963, chap. 4.)

226

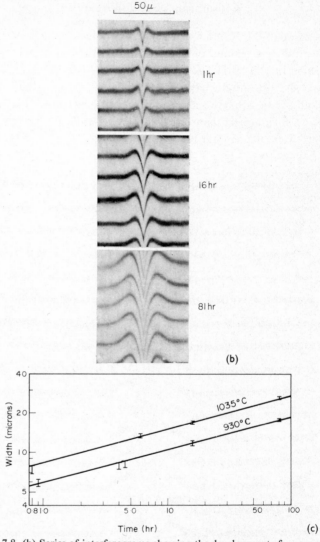

50μ

1hr

16hr

81hr

(b)

FIG. 7.8. (b) Series of interferograms showing the development of a groove in a copper bicrystal after different annealing times at 930°C in hydrogen. The shape of the groove is characteristic of that produced by a diffusive mechanism of mass transport. The variation of the width with time is shown in (c) for two temperatures. The slope of the plot of log (width) versus log (time) is close to 0.25 as predicted by eqn. (7.39). (Figures 7.8(b) and (c) from W. W. Mullins and P. G. Shewman, *Acta Met.* **7**, 163 (1959).)

TABLE 7.2. SOME RESULTS ON SURFACE DIFFUSION FROM MASS TRANSPORT STUDIES

	Surface diffusion activation energy (eV)†	D_0 for surface diffusion (cm²/sec)	Atmosphere	Temperature range (T/T_m)	Method
(i) F.C.C. Metals					
Au (100)	2.35 (1.82)	4×10^5	H_2	0.7–0.995	(a)
Cu (100)	2.0 (2.16)	4.2×10^3	H_2	0.81–0.95	(b)
Cu	0.57 (2.16)		$<10^{-10}$ torr	~ 0.3	(c)
Ir (111)	2.3			0.65–0.8	(c)
Ni (100)	1.5 (2.96)	2.6	$<10^{-8}$ torr	0.7–0.9	(a)
(110) [001]	1.7 (2.96)	12.8			
(110) [110]	1.8 (2.96)	23.9			
Ni	0.91 (2.96)		$<10^{-10}$ torr	0.3–0.44	(c)
Pd	0.91 (2.77)		$<10^{-10}$ torr	~ 0.3	(c)
Pt	1.3 (2.97)		$<10^{-10}$ torr	0.27–0.4	(c)
Rh (111)	1.78		$<10^{-10}$ torr	0.5–0.66	(c)
(ii) B.C.C. Metals					
Fe	2.61 (2.49)	5×10^5	H_2	0.58–0.65	(a)
Mo	2.3 (4.01)		10^{-10} torr	0.6	(c)
Mo (110)*	3.0 (4.01)	4×10^2		0.76–0.83	(a)
Mo (100)*	4.17 (4.01)	2.9×10^4		0.66–0.83	(a)
Ta	2.61 (4.29)		10^{-10} torr		(c)
W	2.96 (6.65)	0.5	10^{-10} torr	0.51–0.6	(c)
W (100)*	5.57 (6.65)	7.6×10^4		0.67–0.86	(a)

(a) Flattening of surface corrugations.
(b) Development of grain boundary grooves.
(c) Field emission studies.

(References to most of the individual experimental measurements may be found in *Surfaces and Interfaces I* (1967), p. 302, or *Surface Phenomena of Metals*, p. 13. See bibliography.)

---

* From K. E. Singer and D. Tabor in *Structure and Properties of Metal Surfaces*, CNRS, Paris, 1970.

† The numbers in brackets are the activation energies for volume self-diffusion; the values quoted are from N. L. Peterson, Diffusion in metals, *Solid State Physics* **22**, 409 (1968).

sion activation energy is in most instances an appreciable fraction of that for volume diffusion.

From the discussion of the previous section it is seen that if surface diffusion proceeds by a defect mechanism the activation energy measured by a mass transport technique will include both the formation energy and migration energy of the defect. A combination of adatom diffusion results, for example, with those from mass transport measurements may thus be able to give information on the adatom formation energy if, indeed, that is the main defect which is involved. Unfortunately both types of result have not yet been obtained for the same surface in the same temperature range. In fact, nonlinear plots of D_s versus $1/T$ have been obtained in some cases, and this suggests that different diffusion mechanisms may be operating in different temperature ranges although the nonlinearities could be due to impurities; adsorbed impurities have been found to have a very important and sometimes spectacular effect on the diffusion coefficient. A number of trends in the experimental data have been noted, in particular that the values of the surface self-diffusion coefficients extrapolated to the melting points are approximately the same[†] for metals with the same structure, a correlation which also holds for bulk diffusion coefficients, but there is as yet no generally accepted model to explain these features.

7.4. ADSORPTION ON SURFACES

Adsorption is a necessary step in numerous reactions at solid surfaces. Heterogeneous catalysis, which is extremely important in biological systems and in many industrial processes, involves reactions among adsorbed molecules which are believed to assume configurations at the surface in which they are particularly reactive. Adsorption is the first step in the growth of oriented thin films (epitaxial growth) and the presence of adsorbed species can strongly influence many phys-

---

[†] For face-centered cubic metals the average value of D_s found by extrapolating experimental results to the melting temperature is $\sim 3 \times 10^{-4}$ cm²/sec.

ical properties of surfaces such as their electron emission characteristics. It is important in vacuum processing and vacuum technology and is directly used in sublimation and cryogenic sorption pumps. To achieve ultra high vacuum it is usually necessary to remove most of the adsorbed gases from the walls of a system by baking at a fairly high temperature ($\gtrsim 250°C$); otherwise the gas slowly desorbs at normal temperatures and in a 1 liter vacuum chamber the release of one monolayer of gas from the walls could conceivably produce a pressure rise of greater than 10^{-3} mmHg (torr).

The simplest situation in adsorption is that of a single atom interacting with a crystal surface. As it approaches the surface the electronic distribution of the atom and of the crystal surface will be changed, and this may result in binding of the atom to the crystal and the creation of an additional surface dipole. A first principles treatment of the problem would ideally yield the new sets of electron energy levels, from which the binding energy could be evaluated, as well as the new wave functions which would give the charge distribution and the change in dipole moment of the surface. Of course, as the coverage increases the adsorbed atoms on a surface cannot be considered isolated, and there may be significant interactions within the adsorbed layer; these will lead to changes with coverage of the average binding energy and dipole moment associated with each adsorbed atom. The nature of the interactions, as well as the mobility of the adsorbate on the surface, will determine whether the adsorption takes place heterogeneously with the formation of islands (or nuclei) with well-defined structure or simply occurs uniformly over the entire surface. Adsorption may lead to dissociation of molecules into individual atoms, ions, or radicals, and it appears that the increased reactivity of molecules at surfaces is related to this dissociation.

While practical situations usually involve polycrystals and gases at relatively high pressures, the fundamentals of the phenomenon of adsorption are most easily studied at relatively low pressures using single crystal surfaces. We shall confine the present discussion to the case of single crystals although a great many laboratory studies of gas adsorption on solids have in fact used polycrystals in the form of finely divided powders, polycrystalline wires, or evaporated films. With materials in these forms the area of surface studied can be made very large so that reasonably large quantities of gas are involved. One of the disadvantages in using single crystals for adsorption studies is that the surface areas are generally small, and this usually leads to difficulties in determining the adsorbate coverage.

7.4.1. *Origin of the Binding Energy in Adsorption*

Molecules in the gas phase have three degrees of translational freedom and perhaps also internal degrees of freedom, vibration, and rotation. In the adsorbed state they will generally execute only vibrational motion (although two-dimensional translation is possible as was indicated in section 7.2) and molecular rotations will be "quenched" to some extent. It is thus clear that the molar entropy to be associated with the adsorbed state will usually be substantially less than that of the gaseous state. In order for adsorption to occur spontaneously there must therefore be a reduction of internal energy or, in other words, a binding energy associated with the adsorbed state.

The potential energy of interaction of a single molecule with a surface as a function of their separation may be represented schematically as in Fig. 7.9; the parameters associated with the equilibrium point and the detailed shape of the potential curve will, of course, depend on the particular molecule and crystal surface involved. The interaction may be discussed qualitatively in terms of the electron energy level scheme of the molecule and the energy bands of the solid. When the two systems are sufficiently close that wave functions start to overlap, there will be some shifting of the energy levels and changes in occupation. We should expect the low-lying levels of the adsorbing molecule to be relatively unperturbed by interaction with the solid since the wave functions which correspond to these states are so localized and are screened by the outer or valence electrons. The levels associated with the outer electrons may, however, undergo substantial shifts. The exact solution for the states of the composite system, crystal plus adsorbed molecule, however, has not yet been obtained

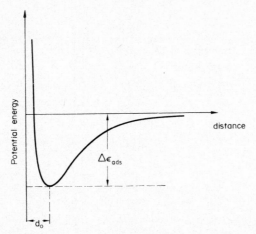

FIG. 7.9. Schematic potential energy versus distance curve for an atom near a planar crystal surface. The minimum corresponds to the equilibrium position for the adsorbed atom at distance d_0 and with binding energy $\Delta\varepsilon_{ads}$. With a real surface the potential energy distance curve will depend on the coordinates in the plane of the surface.

although a substantial amount of research is being devoted to this problem.

We can illustrate some special cases with reference to the diagrams of Fig. 7.10 which refer to the interaction of an atom with a metal. In case (a) both the affinity and ionization levels of the adsorbing atom lie above the Fermi energy of the metal so that the probable direction of electron transfer is from the adsorbate to the metal. This situation is expected to be relatively rare since work functions are generally $\lesssim 5$ eV; it would lead to the formation of an adsorbed positive ion. This may be approximately the situation in the adsorption of alkali atoms on refractory metals. In (b) the affinity level of the atom lies below the Fermi energy and hence would be occupied to produce a negatively charged adsorbate. Case (c) probably represents the most common situation, i.e. $A < \Phi < I$, so that there is no obvious direction of charge transfer and the binding will be due primarily to shifting of energy levels rather than to a change in occupation of pre-existing levels.

16*

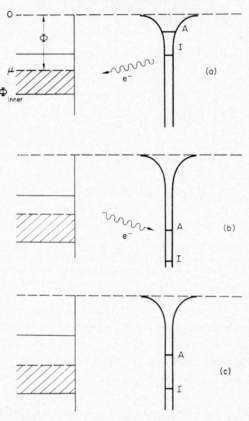

FIG. 7.10. Possible positions of the energy levels of an adsorbing atom relative to the conduction band and Fermi level μ of the metal. The energies A
and I below the vacuum level correspond to the affinity and first ionization
levels of the atom. Assuming that no shifting of levels occurs, the direction
of electron transfer would be from the atom to the metal in case (a), and in
the opposite direction in case (b). In (c) there is no obvious direction of
charge transfer. In a real case the binding will be due to a shifting of the
energy levels as well as to a change in occupancy.

With semiconductors there should be a wider range of possibilities
because of possible electron transfer between the valence and conduction bands of the solid and the affinity and ionization levels of the
adsorbing atom. With the elemental semiconductors, silicon and

germanium, the Fermi level can be varied in quite a predictable manner by suitable doping, and one might be led to the conclusion that the occupancy of the various levels—and hence the adsorbate binding energy—could be changed in a controlled way. However, it appears that with these semiconductors the density of surface states may be sufficiently large that the Fermi level essentially always coincides with the surface states; the position of the Fermi level relative to the bulk bands is unimportant. The adsorbate then interacts with the solid primarily through the intrinsic surface states and essentially does not "see" the underlying band structure.[†]

It has become conventional to distinguish two types of adsorption—*chemisorption* and *physisorption*. Which of these types is exhibited by a particular combination of molecule and surface is usually decided on the basis of the magnitude of the experimental heat of adsorption. While there is considerable spread in values within each category, physical adsorption usually produces a heat of less than about 0.25 eV; heats of chemisorption range from about this value up to several electron volts. The latter are comparable to the energies involved in the formation of normal chemical bonds and, indeed, the term chemisorption is generally used to describe adsorption situations in which there is strong ionic or covalent bonding. On the other hand, the term physisorption (physical adsorption) is used to describe situations where the bonding is believed to be principally due to dispersive interactions and, in fact, the more explicit terminology, *van der Waals adsorption*, is frequently used. It should of course be recognized that the van der Waals interaction will still be operative in cases of chemisorption and, indeed, may make a significant contribution to the total binding energy of the adsorbate to the surface. With inert gases adsorbed on surfaces, the van der Waals interaction is the primary source of binding. Table 7.3 gives examples of the heats of adsorption for a number of cases of chemisorption. In physisorption the adsorbing molecule is more important in determining the magnitude of the

[†] The concept of the "isolated surface" in semiconductors is discussed by T. Wolkenstein in *Adv. in Catalysis* **12,** 189 (1960).

heat with the substrate playing only a secondary role, while, as shown in the table, in chemisorption a particular molecule may be adsorbed on different substrates with widely different heats of adsorption.

TABLE 7.3. SOME EXPERIMENTAL
HEATS OF ADSORPTION IN CASES OF
CHEMISORPTION

Adsorbate	Material	Heat of adsorption (kcal/mole)[a]
O_2	W	194
O_2	Mo	172
O_2	Rh	76
O_2	Pt	70
N_2	W	106
N_2	Ni	10
H_2	W	45
H_2	Mo	40
H_2	Fe	32
H_2	Ni	30
H_2	Rh	26
CO	Ni	35

[a] 1 eV/molecule = 23.05 kcal/mole.

Data from A. W. Adamson, *Physical Chemistry of Surfaces*, 2nd edn., Interscience, 1967, p. 670, and G. Ehrlich in *Metal Surfaces*, American Society for Metals Seminar, 1963, p. 222.

7.4.2. *Thermodynamics of Adsorption*

If we consider a gas in contact with a solid surface as a two-phase[†] one-component system, i.e. unadsorbed gas and adsorbed gas with the substrate being inert, the equilibrium between the two phases can be expressed by a Clapeyron type of equation. Thus for any small changes in the pressure and temperature at fixed number of adsorbed

† In this section we use the idea of a distinct surface phase as discussed on p. l.

moles n_s, the corresponding changes in the partial molar Gibbs free energies of the two phases must be equal at equilibrium, i.e.

$$- \bar{S}_g \, dT + \bar{V}_g \, dP = - \bar{S}_s \, dT + \bar{V}_s \, dP$$

or

$$\left(\frac{\partial P}{\partial T} \right)_{n_s} = \frac{\bar{S}_g - \bar{S}_s}{\bar{V}_g - \bar{V}_s} = \frac{\bar{H}_g - \bar{H}_s}{T(\bar{V}_g - \bar{V}_s)}, \tag{7.42}$$

which leads to

$$\frac{1}{P} \left(\frac{\partial P}{\partial T} \right)_{n_s} = \frac{\Delta \bar{H}_{\text{ads}}}{RT^2} \tag{7.43}$$

when we assume ideal gas behavior and neglect the volume of the adsorbed gas. The partial molar quantity $\Delta \bar{H}_{\text{ads}}$ is termed the *isosteric heat of adsorption* since it refers to constant gas coverage on the surface. $\Delta \bar{H}_{\text{ads}}$ will in general depend on the coverage n_s and temperature T. The heat of adsorption and energy of adsorption $\Delta \bar{E}_{\text{ads}}$ $[=(\bar{E}_g - \bar{E}_s)]$ are related by

$$\Delta \bar{H}_{\text{ads}} = \Delta \bar{E}_{\text{ads}} + (P\bar{V}_g - P\bar{V}_s) \approx \Delta \bar{E}_{\text{ads}} + RT \tag{7.44}$$

when we neglect the volume of the adsorbed phase provided the gas pressure is not too high for deviations from ideality in the gas phase to be significant. The difference between the two quantities (~ 0.6 kcal/mole at $300°K$) may be important in physical adsorption, but it is usually negligible in systems exhibiting chemisorption.

The heat of chemisorption can be measured directly by calorimetric techniques or through measurements of the rates of desorption (see below) promoted by heating the surface or by irradiating with electrons, photons or ions. Heats of adsorption can also be deduced for reversible systems by determining the equilibrium coverage θ as a function of gas pressure at various temperatures, i.e. the *adsorption isotherms*. From such plots isosteres (plots of the pressure and temperature conditions necessary to maintain fixed coverages) can be derived and interpreted according to eqn. (7.43).

We have already encountered adsorption isotherms in section 2.2 in connection with the Gibbs formulation of surface excess quantities.

A number of different forms of isotherms are possible. For chemisorption on rather uniform surfaces the isotherms frequently encountered are roughly of the *Langmuir* type shown schematically in Fig. 7.11, while for more heterogeneous adsorbents and multilayer adsorption, more complicated isotherms[†] are obtained. It is of interest to see what type of assumptions lead to the Langmuir isotherm.

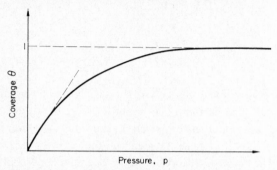

FIG. 7.11. Langmuir type of adsorption isotherm for monolayer adsorption. At sufficiently low pressures the coverage is proportional to the pressure (the so-called Henry's law region) while at higher pressures the coverage becomes saturated. More complex adsorption isotherms are associated with heterogeneous adsorbents, the formation of multilayers of adsorbed atoms, and lateral interactions among the adsorbed atoms.

The potential energy of interaction of a molecule with a crystalline surface will vary with the normal distance from the surface in the manner indicated in Fig. 7.9, and it will also have a doubly periodic variation with displacement parallel to the surface (Fig. 7.3). We will suppose that the magnitude of the potential variation parallel to the surface is sufficiently large in comparison to kT that the adsorbed molecules will be localized most of the time and may be considered to have three independent degrees of vibrational freedom. The partition function for a single adsorbed molecule $q(T)$ is then

$$q(T) = q_x q_y q_z \exp\left(-\frac{U_0}{kT}\right), \qquad (7.45)$$

[†] Various types of isotherms are discussed in the book *The Solid–Gas Interface* (ed. E. A. Flood), Dekker, New York, 1966.

where U_0 is the potential energy of the molecule in an equilibrium adsorption site on the surface relative to that of a molecule at rest at infinite separation from the crystal, and q_x, q_y, and q_z are the individual one-dimensional harmonic oscillator partition functions. We suppose there are M' equivalent sites available to the adsorbing molecules and that at a particular temperature and gas pressure the equilibrium number of molecules on the surface is N. We assume further that there is *no interaction between adsorbed molecules* even when they are in neighboring sites. The partition function for the system of adsorbed molecules $Q(N, M', T)$ is then given by

$$Q(N, M', T) = \frac{M'! \, q(T)^N}{N! \, (M'-N)!}. \tag{7.46}$$

The equilibrium situation is that in which the chemical potential is the same in the adsorbed phase as in the gaseous phase. For the adsorbed phase the chemical potential (defined by $\mu = (\partial F/\partial N)_{T, M'}$) is related to the partition function by

$$\mu = -kT \left(\frac{\partial \ln Q}{\partial N} \right)_{T, M'}. \tag{7.47}$$

From eqn. (7.46) (with Stirling's approximation)

$$\mu = kT \ln \frac{N}{(M'-N) \, q(T)} = kT \ln \frac{\theta}{(1-\theta) q(T)}, \tag{7.48}$$

where $\theta = N/M'$ is the fractional occupancy of the available surface sites and is termed the *surface coverage*. For the gaseous phase, with the assumption of ideality, the chemical potential is given by

$$\mu = \mu_0(T) + kT \ln p, \tag{7.49}$$

where $\mu_0(T)$ is the standard state chemical potential (generally chosen as the value of μ at a gas pressure of 1 atm.). Combining eqns. (7.48) and (7.49) we then have the relationship between θ and p at equilibrium,

$$\theta = \frac{f(T)p}{1+f(T)p}, \tag{7.50}$$

where $f(T) = q(T) \exp (\mu_0(T)/kT)$. Equation (7.50) has the form of the so-called *Langmuir isotherm* initially derived by Langmuir by kinetic arguments; at very low pressures $\theta \propto p$, while at high pressures the surface becomes saturated, i.e. $\theta \to 1$. Figure 7.12(a) shows an example of a set of isotherms for the adsorption of sulfur on the (111) surface of silver. These show the saturation effect, but a detailed examination shows that they depart significantly from the shape predicted by eqn. (7.50) probably due to lateral interactions among the adsorbed atoms. Equation (7.50) is strictly valid only when the adsorbate is immobile

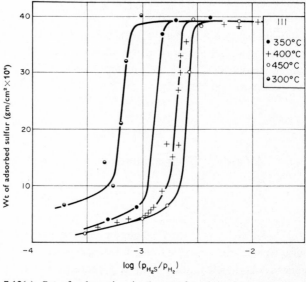

FIG. 7.12(a). Set of adsorption isotherms for the adsorption of sulfur on the (111) surface of silver. The amount of adsorbed sulfur was determined by a radioactive tracer method for various values of the ratio of the partial pressures of H_2S and H_2 in contact with the surface. The effective partial pressure of sulfur is given by $\log p_{S_2} = \log K + 2 \log (p_{H_2S}/p_{H_2})$, where K is the equilibrium constant for the dissociation of H_2S. Note that the pressure scale is logarithmic. The isotherms show saturation adsorption at one monolayer of sulfur ($\sim 40 \times 10^{-9}$ gm/cm²) but depart somewhat from the Langmuir shape probably due to interactions among the adsorbed atoms.

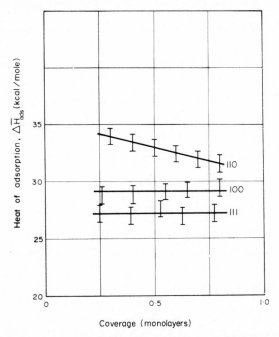

Coverage (monolayers)

FIG. 7.12(b). Heat of adsorption of sulfur on different faces of silver crystals as a function of coverage. It is essentially constant for (111) and (100) up to one monolayer but decreases significantly with coverage for (110). (From J. Bénard, J. Oudar, and F. Cabané-Brouty, *Surface Sci.* **3**, 359 (1965).)

and there is no lateral interaction between adsorbed molecules. From the discussion of Chapter 4 it will be clear that the binding energy of an adsorbed atom may vary with the orientation of the surface (Fig. 7.12(b)) and that even on a particular surface there will in general be a variety of binding sites for adsorbed atoms with a range of energies of adsorption. It has been suggested that heterogeneity and lateral interaction may have compensating influences on the shape of the isotherm.

It is worth noting that in most systems which exhibit chemisorption the equilibrium coverage at room temperature is very close to unity even at very low pressures. For example, if we take $U_0 \approx -2\,\text{eV}$, assume the classical limit $(kT/h\nu)$ for the harmonic oscillator partition functions with the vibrational frequencies taken to be about 10^{12} sec^{-1}, set μ_0/kT equal to

$$\ln\left\{\left(\frac{h^2}{2\pi m}\right)^{3/2}\left(\frac{1}{kT}\right)^{5/2}\right\},$$

the value appropriate to a monatomic gas, we find from (7.50) that the gas pressure required to produce an equilibrium coverage of about 0.9 monolayer at room temperature is only of the order of 10^{-24} torr; for $U_0 \approx -1$ eV the corresponding pressure is about 10^{-7} torr.

7.4.3. *Structure of Adsorbed Layers on Single Crystals*

Recently the techniques of electron diffraction and field electron and ion microscopy have been used to obtain information on the detailed arrangement of adsorbed atoms on surfaces, their mobilities, and the mechanisms of adsorption. The diffraction work has demonstrated the frequent occurrence of long range order within the adsorbed layer with periodicities related to those of the substrate. In some cases the primitive vectors characterizing the adsorbed layer are several times those of the substrate surface, indicating the presence of long range interactions between adsorbate molecules. In many cases adsorbed layers do not form homogeneously but nucleation occurs heterogeneously on the surface, and the proportions of different structures change as the average coverage increases. We can illustrate some features that are common to many of these studies[†] by considering a particular system, the interaction of oxygen with tungsten. We consider adsorption of oxygen on the (112) surface of tungsten at room temperature. This system has been studied extensively and exhibits some rather interesting features. Figure 7.13(a) is a hard sphere model of a (112) surface of a body-centered cubic metal showing the close-

---

[†] Much of the research work on the structure of absorbed layers was initiated by H. E. Farnsworth and by L. H. Germer and their coworkers. Comprehensive surveys of the observations are given in the article by J. W. May cited in the bibliography.

packed rows of atoms separated by relatively large distances. Figure 7.13(b) is the corresponding low energy electron diffraction pattern taken at normal incidence. As oxygen is allowed to adsorb on the surface, extra diffracted beams appear at the positions of type $(h + \frac{1}{2}, k)$ indicating the presence of an additional structure with a double periodicity in the $[11\bar{1}]$ direction. It will be noted that the extra diffraction beams are spread out in the $[1\bar{1}0]$ direction at low coverages (Fig. 7.13(c)) and become sharper as the coverage approaches $1/2$ monolayer (Fig. 7.13(d)). The probable explanation of this observation rests on the fact that for this particular adsorbed structure there are two possible positions of the primitive cell which produce out of phase contributions to the $1/2$ order beams. Islands in the two positions are referred to as *antiphase domains* and are possible whenever the adsorbed layer has a repeat distance which is a multiple of that of the substrate. The direction of the streaks in the diffraction pattern indicated the islands are very narrow in the $[1\bar{1}0]$ direction, and this type of morphology appears to be consistent with the expected relative mobilities of the adsorbed atoms/molecules in the two orthogonal directions, i.e. that motion in the $[1\bar{1}0]$ should be much more difficult than in the $[11\bar{1}]$. Figure 7.13(e) and (f) shows further stages in the adsorption process, the formation of the full monolayer structure, and the $3/2$ monolayer structure; the latter corresponds to a double periodicity in the $[1\bar{1}0]$ directions. Numerous examples of such super-lattices can be found in the literature observed either by low energy electron diffraction or reflection high energy electron diffraction.

It has been suggested that adsorption may be accompanied by displacements of the substrate atoms over interatomic distances, a process called *reconstruction*. The evidence in support of this is rather indirect and there does not yet appear to be any system for which there is general agreement that reconstruction has occurred. It is hoped that once a satisfactory theory of low energy electron diffraction has been developed, the details of adsorbed structures will become known. Recently significant progress has been made in the theory of low energy electron diffraction and tentative assignments of unit mesh contents have been made in a few cases for adsorbed layers.

242 PROPERTIES OF CRYSTAL SURFACES

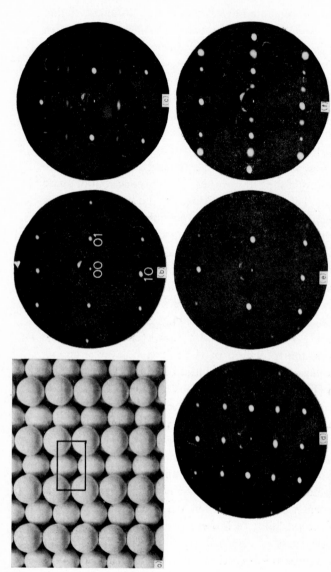

FIG. 7.13. Example of the formation of ordered adsorbed structures. The particular system considered here is that of oxygen interacting with the (112) surface of tungsten. (a) A model of an ideal (112) surface of a body-centered cubic crystal with the surface unit mesh indicated, and (b) the low energy electron diffraction pattern from a clean (112) surface; (c) and (d) show the formation of the ordered $\frac{1}{2}$ monolayer structure; (c) corresponds to a coverage of about 0.35 monolayer and (d) to about 0.5 monolayer. The fact that the additional spots are present from low oxygen coverages and become sharper and more intense as the coverage approaches $\frac{1}{2}$ monolayer suggests that the (2×1) structure exists as islands on the surface whose fractional coverage of the surface increases with average coverage; (e) and (f) correspond to later stages in the adsorption sequence; (e) is the ordered full monolayer structure (1×1), while (f) corresponds to about $\frac{3}{4}$ monolayers (1×2). (Photographs courtesy of C. C. Chang.)

7.4.4. Work Function Changes during Adsorption

Adsorption usually produces a change in the work function of the surface. As can be seen by considering the various contributions to the work function as outlined in section 5.4.3, the quantity which is changed by adsorption is the surface double layer contribution. This change in work function is often expressed in terms of an effective dipole moment μ' associated with each adsorbed molecule and defined by

$$\Delta\Phi = (\Phi - \Phi_0) = 4\pi N' \theta \mu', \qquad (7.51)$$

where Φ_0 and Φ are the work function before and after adsorption, N' is the number of adsorbed atoms per unit area which would constitute a monolayer, and θ is the fractional coverage. The dipole moment μ' may be partly due to a permanent dipole of the adsorbing molecule (as, for example, in the case of adsorbed CO) or to the redistribution of charge in the atom–metal system; in the latter case

$$\mu' = \frac{1}{N'\theta} \int_v \mathbf{r} \, \Delta\varrho(\mathbf{r}) \, dv, \qquad (7.52)$$

where $\Delta\varrho(\mathbf{r})$ is the change in charge density at $\mathbf{r}$ due to the adsorption, the integration being taken over the charge in a cylinder of unit cross-sectional area extending a large distance on either side of the surface. As we have already noted, there may be a significant interaction between individual adsorbed atoms on the surface with the result that μ' is in general a function of the coverage θ. An example of the variation of $\Delta\Phi$ with θ for the (112) surface of tungsten is shown in Fig. 7.14(a). In this case $\Delta\Phi$ increases smoothly from the clean surface value, shows a slight minimum at about one monolayer and then increases only very slowly with further adsorption. Figure 7.14(b) shows a somewhat different result for the adsorption of sodium on a (100) surface of nickel. At low coverages the work function decreases rapidly, reaches a minimum, and then approaches the value characteristic of a sodium surface. The corresponding variations of the effective

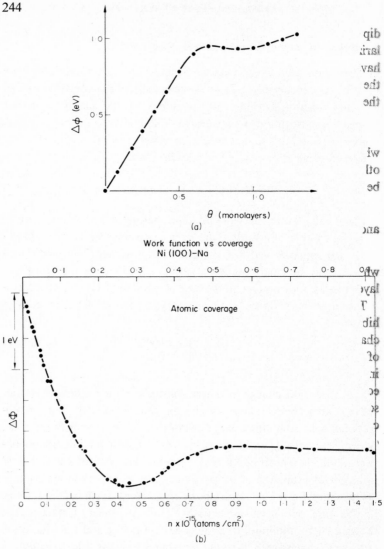

FIG. 7.14. (a) Change in work function with coverage of the (112) surface of tungsten with oxygen. The initial dipole moment using eqn. (7.51) is ~0.4 D. (From J. C. Tracy and J. M. Blakely in *The Structure and Chemistry of Solid Surfaces* (ed. G. Somorjai), Wiley, 1969.) (b) The corresponding set of curves for the adsorption of sodium on a (100) nickel surface. In this case the work function has a pronounced minimum at about 0.25 monolayer. The initial dipole moment obtained using eqn. (7.51) is ~3.5 D. (The unit of dipole moment is the debye (D) where $1 D = 10^{18}$ e.s.u.) (Figure 7.14(b) courtesy of R. L. Gerlach.)

ole moment μ' with θ are often expressed in terms of mutual depo-
arization of the adsorbate atoms or molecules. For example, if we
have an array of dipoles of polarizability α and suppose that each in
he isolated adsorbed state had moment μ_0', then at any finite coverage
he dipole moment of a particular adsorbed atom is

$$\mu_i' = \mu_0' - \alpha E_D^i, \tag{7.53}$$

where E_D^i is the effective depolarizing field at dipole i due to all the
her dipoles. If all the dipoles may be regarded as identical, as will
be the case for a completely ordered array,

$$E_D = \beta\mu' \tag{7.54}$$

and hence

$$\mu' = \mu_0'/(1+\alpha\beta), \tag{7.55}$$

where β is a parameter determined by the geometry of the adsorbed
ayer and the nature of the screening by the metal electrons.

The changes in work function accompanying adsorption are exhibited most dramatically in either the thermionic or field emission
characteristics of solids. This is, of course, due to the strong dependence
of the emission current on the work function of the solid expressed
in the Richardson–Dushman eqn. (6.11) and the Fowler–Nordheim
eqn. (6.1). It is this sensitivity which makes the field emission microscope very useful for the study of adsorption; Fig. 7.15 shows the
changes in emission patterns from a tip accompanying the adsorption
of oxygen on nickel. Because of such dramatic changes in emission it
is possible to study the structure of extremely small quantities of
adsorbed gas and even detect effects due to single adsorbed atoms.
The rates of migration of adsorbates over field emitter surfaces have
also been studied quite extensively with field emission methods for a
fairly wide variety of systems.

7.4.5. Kinetics of Adsorption and Desorption

Corresponding to prescribed conditions of temperature and pressure
of a gas in contact with a surface we have seen that there will be a par-

O$_2$ on nickel

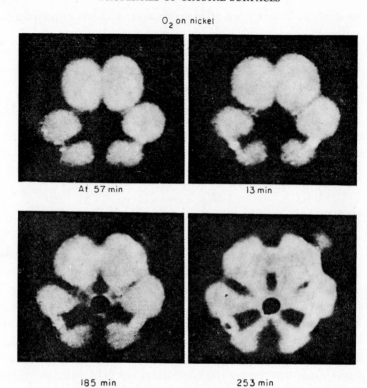

At 57 min   13 min

185 min   253 min

FIG. 7.15. Series of field emission patterns from a nickel specimen at various
stages during the interaction with oxygen showing dramatic changes in the
emission characteristics with adsorption. The oxygen pressure was about
10^{-8} – 10^{-9} mmHg. (From G. Ehrlich in *Metal Surfaces*, American Society
of Metals, Cleveland, 1963.)

ticular equilibrium coverage of gas molecules on the surface and these
will assume some stable structure and morphology. The equilibrium
condition is uniquely defined and is quite independent of how the
adsorption takes place. However, the kinetics of the adsorption process
are usually of considerable interest in practice and very often information on the equilibrium structure, binding energy, and morphology
can be obtained by studying the kinetics of its formation and its
break-up.

The adsorption of diatomic molecules on a surface to form an ordered atomic array may be supposed to occur in a number of consecutive steps: (a) the collision of the molecules with the surface with transfer of their momentum to the crystalline lattice, (b) migration of the molecule over the lattice, and (c) dissociation and incorporation into the equilibrium structure. The rate of each of these processes is expected to be temperature dependent and each may be the step which limits the rate of the overall reaction. Adsorption processes in which the rate increases exponentially with temperature are said to be *activated*.

The rate of adsorption is usually expressed in terms of a quantity called the *sticking coefficient* (or sticking probability) S which is in general a function of temperature and coverage θ for any given combination of gas and crystalline surface. It is defined as follows. The number of molecules I striking a surface, which is in contact with a gas at pressure p, per unit area per unit time, is from the kinetic theory of gases

$$I = p/(2\pi mkT)^{1/2},$$

where m is the mass of a gas molecule; the total number of molecules incident in time t, i.e. the *exposure E*, is $\int I\, dt$. The sticking coefficient S at any fractional coverage θ is then defined as

$$S(\theta, T) = N'\left(\frac{\partial \theta}{\partial E}\right), \tag{7.56}$$

where N' is the number of molecules in one monolayer. Very often the quantity determined from experiment is a mean sticking coefficient $\bar{S}$ relating the coverage to the total exposure, i.e.

$$\bar{S} = N'\left(\frac{\theta}{E}\right). \tag{7.57}$$

Some possible variations of coverage with exposure for the case where the maximum amount of adsorbed gas constitutes one mono-

17*

layer are shown in Fig. 7.16 with the corresponding variations of S with E. If the adsorbate has essentially no mobility on the surface but is chemisorbed only if it impinges directly on a suitable unoccupied site the sticking probability will vary as $(1-\theta)$ for monolayer adsorption, this being the probability that any particular adsorption site is vacant. In cases where the adsorbate is capable of diffusive motion on the surface, the dependence of S on θ may depend appre-

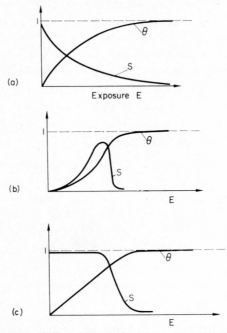

FIG. 7.16. Some possible variations of coverage θ and sticking coefficient S with exposure E. In (a) the sticking coefficient varies as $(1-\theta)$; this would be the case if adsorption occurred only if the incident gas atom impinged directly on an unoccupied adsorption site. The variation of S and θ with E depicted in (b) could arise if adsorption occurs by the process of nucleation and growth. The sticking coefficient would increase as the number of empty sites at the edges of growing nuclei increases; the fall off at higher exposures would be due to impingement of neighboring nuclei. (c) corresponds to the situation where the adsorbing species is very mobile on the surface and essentially all physisorbed atoms or molecules become chemisorbed.

ciably on the morphology of the adsorbate layer. To illustrate this we may consider the example where a molecule is adsorbed in an intermediate weakly bound physisorbed state (or so-called precursor state) and can migrate over the surface with diffusivity D. There will be some reasonable probability of re-evaporation before dissociation occurs, and if τ_e is the lifetime for re-evaporation, the root mean square displacement R before desorption is $\sqrt{(2D\tau_e)}$. The surface diffusivity appropriate to the precursor state increases exponentially with temperature being proportional to $\exp(-Q_s/RT)$, where Q_s is the activation energy for surface diffusion of the adsorbed molecule, while the mean lifetime on the surface will decrease with increasing temperature being proportional to $\exp(\Delta \bar{H}_{ads}/RT)$. The root mean square displacement R is then given by

$$R \propto \exp\left\{\frac{(\Delta \bar{H}_{ads} - Q_s)}{RT}\right\}, \qquad (7.58)$$

which will in general decrease with increasing temperature since the energy of desorption is expected to be greater than the energy variation encountered during diffusion over the surface.[†] If the diffusion length R is large compared to the average separation of sites on the surface at which chemisorption occurs then nearly all molecules which enter the precursor state will become chemisorbed. Under these conditions the chemisorption rate is determined by the rate of physisorption which may be essentially coverage independent. On the other hand, if the diffusion length is smaller than the separation of potential chemisorption sites the sticking coefficient or chemisorption rate may be very dependent on coverage. In the latter case if the chemisorbed atoms are arranged in islands on the surface only those molecules initially physisorbed within a distance R of the edge of an island will

---

[†] An order of magnitude estimate for R may be obtained as follows. The mean lifetime for desorption may be written as $1/\tau_e \approx \nu \exp(+U_0/kT)$ by analogy with eqn. (7.10) where U_0 is defined in eqn. (7.45). Combining with eqn. (7.11) for diffusion between adsorption sites gives $D\tau \approx l^2 \exp[(-2U_0 - V_0)/2kT]$; setting $U_0 \approx -0.25$ eV, $V_0 \approx 0.1$ eV, and taking l to be one interatomic distance would give $R = \sqrt{(2D\tau_e)} \approx 50$ Å at room temperature.

be chemisorbed before desorption. This leads to the concept of an *active zone* surrounding each nucleus or island. If we assume the initial nuclei to be circular in shape the total area within the active zones will vary approximately as $\sqrt{\theta}$, the dimension of the total island perimeter, until the islands impinge on each other. Thus the chemisorption rate will vary as $\sqrt{\theta}$ following nucleation and then will tend to zero as the coverage approaches unity.

The measurement of the rate at which gas is removed from a surface can be used to give a considerable amount of information on the binding in the adsorbed state. Figure 7.17 shows the amount of nitrogen

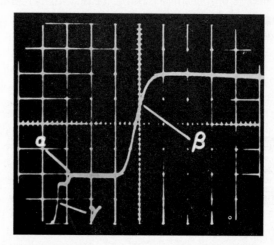

FIG. 7.17. Flash desorption of nitrogen from a polycrystalline tungsten filament. The ordinate is the gas density and the abscissa the time during heating. The three separate sharp increases labeled γ, α, β in gas density (pressure) indicate three different states of binding of nitrogen to tungsten surfaces. (From G. Ehrlich, *J. Appl. Phys.* **32**, 4 (1961).)

gas released from a polycrystalline tungsten filament as a function of time as the temperature is continuously increased. It will be noted that the desorption occurs in fairly discrete steps so that the rate of desorption exhibits peaks at particular temperatures for a given rate of temperature increase. These thermally stimulated desorption peaks indicate the existence of different binding states of nitrogen on the tungsten surface perhaps associated with different kinds of surface sites. However, it is not clear whether adsorbed molecules exist in these states in equilibrium or whether they are states which exist only during the desorption process.

Adsorbed molecules can also be removed from a surface through interaction with a beam of electrons, ions, or photons. The phenomenon of electron stimulated desorption has in fact been studied quite extensively[†] in an attempt to obtain information on adsorbate binding energies. The details of the mechanisms by which the desorption occurs are not yet completely known, but it is clear that the process is not simply one of momentum transfer from the incident electron to the adsorbed molecule but that desorption probably results from electronic transitions occurring in the adsorbate-crystal system. In several cases it has been possible to obtain values for the energies of adsorption by determining the electron threshold energy for desorption; different binding states for molecules on some surfaces have also been found. In addition to direct desorption the incident electrons can alter the bonding of adsorbed molecules, produce dissociation or the desorption of molecular fragments. Measurements of the electron energy loss spectra, emitted ion or atom energy distribution and desorption threshold energies are presently being made and a synthesis of these data will hopefully yield details of adsorbate-crystal bonding.

BIBLIOGRAPHY

Adsorption

BÉNARD, J., OUDAR, J., and CABANÉ-BROUTY, Contribution à l'étude de l'adsorption chimique réversible du soufre sur l'argent, *Surface Sci.* **3**, 359 (1965).
BENNETT, A. J., McCARROLL, B., MESSMER, R. P., A molecular orbital approach to chemisorption, *Surface Sci.* **24**, 191 (1971).
CLARK, A., *The Theory of Adsorption and Catalysis*, Academic Press, 1970.
DRAUGLIS, E., GRETZ, R. D., and JAFFEE, R. I. (eds.), *Molecular Processes on Solid Surfaces* (part 4), McGraw-Hill, 1969.

[†] A comprehensive review of the experimental and theoretical work on electron stimulated description is given in the article by T. E. Madey and J. T. Yates cited in the bibliography.

EHRLICH, G., Adsorption and surface structure, in *Metal Surfaces*, American Society for Metals, Cleveland, 1963.

FLOOD, E. A., (ed.), *The Solid–Gas Interface*, Dekker, New York, 1967.

HAGSTRUM, H. D., and BECKER, G. E., Orbital energy spectra of electrons in chemisorption bonds: O, S, Se on Ni (100), *J. Chem. Phys.* **54**, 1015 (1971).

HAYWARD, D. O., and TRAPNELL, B. M. W., *Chemisorption*, 2nd edn., Butterworths, London, 1964.

MADEY, T. E. and YATES, J. T., Electron stimulated desorption as a tool for studies of chemisorption,. *J. Vac. Sci. Tech.* **8**, 525 (1971).

MAY, J. W., Discovery of surface phases by low energy electron diffraction, *Advances in Catalysis* **21**, 151 (1970).

REDHEAD, P. A., HOBSON, J. P., KORNELSEN, E. V., *The Physical Basis of Ultrahigh Vacuum*, Chapman & Hall, 1968.

RIDEAL, E. K., *Concepts in Catalysis*, Academic Press, 1968.

ROBERTSON, A. J. B., *Catalysis of Gas Reactions by Metals*, Logos Press Ltd., 1970.

SALTSBURG, H., SMITH, J. N., and ROGERS, M. (eds.), *Fundamentals of Gas–Solid Interactions*, Academic Press, 1967.

SCHRIEFFER, J. R., Theory of chemisorption, *J. Vac. Sci. Tech.* **9**, 561 (1972).

SLATER, J. C. and JOHNSON, K. H., Self-consistent field X α cluster method for polyatomic molecules and solids, *Phys. Rev.* B, **5**, 844 (1972).

Adsorption et Croissance Cristalline, CNRS, Paris, 1965.

Advances in Catalysis, Vol. 21, Academic Press, 1970.

Electronic Phenomena in Chemisorption and Catalysis on Semi-conductors (ed. K. Hauffe and Th. Wolkenstein), Walter de Gruyter, Berlin, 1969.

Sorption–desorption phenomena in high vacuum, Supplement to *Nuovo Cimento* **5**, (2) (1967).

The role of the adsorbed state in heterogeneous catalysis, *Disc. Faraday Soc.*, No. 41, 1966.

Surface Diffusion

BLAKELY, J. M., Surface diffusion, *Progress in Materials Science*, **10**, (4) (1963).

DE BOER, J. H., Mobility of molecules along adsorbing surfaces, in *Molecular Processes on Solid Surfaces* (ed. E. Drauglis, R. D. Gretz, and R. I. Jaffee), McGraw-Hill, 1969.

DELAMARE, F., and RHEAD, G. E., Increase in the surface self-diffusion of copper due to the chemisorption of halogens, *Surface Sci.* **28**, 267 (1971).

EHRLICH, G., Atomistics of metal surfaces, in *Surface Phenomena of Metals*, Society of Chemical Industry, Monograph No. 28, 1968.

EHRLICH, G., and HUDDA, F., Atomic view of surface self-diffusion: tungsten on tungsten, *J. Chem. Phys.* **44**, 1039 (1966).

GJOSTEIN, N. A., Surface self-diffusion in metals, in *Surfaces and Interfaces I*, (ed. J. J. Burke, N. L. Reed, and V. Weiss), Syracuse Univ. Press, 1967.

HENRION, J., and RHEAD, G. E., in *Diffusion Processes*, Gordon & Breach, New York, 1970.

HIRTH, J. P., The kinetic and thermodynamic properties of surfaces, in *Energetics in Metallurgical Phenomena*, Vol. 2 (ed. W. M. Mueller), Gordon & Breach, 1965.

KNACKE, O., and STRANSKI, I. N., The mechanism of evaporation, *Progress in Metal Physics* **6**, 181 (1956).

MULLINS, W. W., Solid surface morphologies governed by capillarity, *Metal Surfaces*, American Society for Metals, Cleveland, 1963.

NEUMANN, G. M., and NEUMANN, K., *Surface Diffusion*, (1971).

SOME GENERAL REFERENCES

THERMODYNAMICS

GUGGENHEIM, E. A., *Thermodynamics*, North-Holland, 1949.
HILL, T., *Introduction to Statistical Thermodynamics*, Addison-Wesley, 1960.
SWALIN, R. A., *Thermodynamics of Solids*, Wiley, New York, 1962.

ATOMIC AND SOLID STATE PHYSICS

DEKKER, A. J., *Solid State Physics*, Prentice-Hall, 1957.
KITTEL, C., *Introduction to Solid State Physics*, 4th edn., Wiley, 1971.
RAIMES, S., *The Wave Mechanics of Electrons in Metals*, North-Holland, 1961.
SPROULL, R. L., *Modern Physics*, 2nd edn., Wiley, 1963.

TEXTS ON SURFACES

ADAMSON, A. W., *Physical Chemistry of Surfaces*, 2nd edn., Interscience, 1967.
BIKERMAN, J. J., *Physical Surfaces*, Academic Press, 1970.
FRANKL, D. R., *Electrical Properties of Semi-conductor Surfaces*, Pergamon, 1967.
HAYWARD, D. O., and TRAPNELL, B. M. W., *Chemisorption*, 2nd edn., Butterworths, London, 1964.
MANY, A., GOLDSTEIN, Y., and GROVER, N. B., *Semiconductor Surfaces*, North-Holland, 1965.
SOMORJAI, G. A., *Principles of Surface Chemistry*, Prentice-Hall, 1972.

CONFERENCE BOOKS AND REVIEW SERIES

Fundamental Phenomena in the Materials Sciences (3rd Annual Symposium) (ed. L. J. Bonis, P. L. de Bruyn, J. J. Duga), Plenum Press, 1966.
Fundamentals of Gas–Surface Interactions (ed. H. Saltsburg, J. N. Smith, and M. Rogers), Academic Press, 1967.
Metal Surfaces, American Society for Metals, New York, 1963.

Structure et Propriétés des Surfaces des Solides, C.N.R.S., Paris, 1970.
Molecular Processes on Solid Surfaces (ed. E. Drauglis, R. D. Gretz, R. I. Jaffee), McGraw-Hill, 1969.
Recent Progress in Surface Science (ed. J. F. Danielli, K. G. A. Pankhurst, A. C. Riddiford), Academic Press, from 1964.
Solid State Surface Science (ed. M. Green), Dekker, from 1969.
The Solid–Gas Interface (ed. E. A. Flood), Vols. I and II, Dekker, 1967.
The Structure and Chemistry of Solid Surfaces (ed. G. Somorjai), Wiley, 1969.

INDEX